Peter F Smith & Adrian C Pitts

concepts in practice
energy

Building for the Third Millennium

First published in Great Britain 1997

A CIP catalogue record for this book is available from the British Library.

ISBN 0 7134 7875 6

Printed in Singapore

For the Publishers

B.T. Batsford Ltd
583 Fulham Road
London SW6 5BY

Contents

Case studies

Conversion table

Metric/imperial conversion factors for the units used in this book

To convert from	to	multiply by
C	F	1.8 (and add 32 for absolute values)
mm	inches	0.039
cm	inches	0.39
cm	feet	0.03281
m	feet	3.281
km	miles	0.6215
m^2	ft^2	10.76
km^2	square miles	0.3861
hectares	acres	2.471
m^3	ft^3	35.31
km^3	cubic miles	0.2401
litres	US gallons	0.2642
kg	lb	2.205
m/s	miles per hour	0.6215
pascals (Pa)	inches water gauge	0.004
tonnes	tons	1.016
kg/m^2	lb/ft^2	0.205
kW	btu/hour	3412
kW	horsepower	1.341
kWh	btu	3412
kWh/m^2	btu/ft^2	317.1
W/m^2 C	btu/hour ft^2 F	0.1761
gJ/m	$therms/ft^2$	0.881

Acknowledgements

Thanks are due to many people who have contributed to or helped in the production of this book: Chris Scott for his initial help in compiling information for the case studies; Stuart Craigen for producing a number of drawings, and Steve Sharples and Ian Ward (University of Sheffield) for their discussions and help in obtaining information. The authors would also like to thank Steve Baker (Peter Foggo Associates), Richard Brearley (John Miller and Partners), Sandra Dixon and Nicola Pearsall (University of Northumbria), John Doggart (ECD Energy and Environment), Juliet Wood (ECD Architects), David Emond (RH Partnership), William Gething (Feilden Clegg Architects), Christopher Nash (Nicholas Grimshaw and Partners), Russell Read (Linacre College, Oxford), Susan Roaf (Oxford Brookes University), Huw Turner (Foster and Partners), Chris Twinn of Ove Arup and Partners and Michael Hopkins and Partners, Brenda and Robert Vale, and Robert Weber for their help in compiling individual case studies.Particular thanks are due to David Olivier for help with domestic case studies and Dr Bill Bordass for providing information for the concluding chapters.

Particular thanks are due to those who provided some of the illustrations:

Ove Arup and Partners: Figure 4.39

Peter Cook: Figures 4.23, 4.24 and 4.25

ECD Architects/Energy and Environment: Figures 4.1, 4.3, 4.4, 4.16, 4.17, 4.18, 4.19 and 4.20

Norman Foster and Partners: Figures 4.12, 4.13, 4.4 and 4.15

Michael Hopkins and Partners: Figures 4.36, 4.37 and 4.38

John Linden: Figure 4.26

David Olivier: Figure 3.2

Timothy Soar: Cover photograph, Figures 4.21, 4.22, 4.31, 4.32, 4.34 and 4.35

Robert Stangier,Swiss Federal Laboratories for Materials testing and Research: Figures 3.1 and 3.3

Brenda and Robert Vale: Figures 3.10 and 3.14

Anthony Weller: Figures 4.27, 4.28, 4.29 and 4.30

Figure 1.2 is reproduced with permission from *Nature* (Vol. 347, September 1990, p.139) copyright © 1990 Macmillan Magazines Ltd.

Figure A.1 is derived from *New Scientist*, 26 November 1994, p.6

1. Introduction: A global concern

Design in a changing climate

Much has been written about the merits and techniques of environmentally friendly architecture and sustainability, and there is general agreement that the goal of sustainability is worth striving to accomplish. It is also widely realized that in the industrialized countries, buildings are still the main threat to sustainability because of their heavy reliance – either directly or indirectly – on fossil-based energy.

Later in this chapter, we will describe the serious problems facing humanity, and seek to encourage the creation of buildings which harmonize with nature rather than fight it. Whilst most people, including building designers, are aware of the basic theory behind global warming, what is not so widely understood is the extent of the scientific evidence which underpins it, nor the predictions regarding the impact of climate change. If there is to be a significant shift towards 'biomorphic' architecture, then both building designers and their clients will need to be convinced that it is a matter of extreme importance, for the *present* as well as for future generations. (The term 'biomorphic' has been chosen in preference to the more common 'bioclimatic' because it more accurately describes the connection between the biosphere and the form of buildings.) This is why the authors consider it is appropriate to begin with a summary of the main evidence supporting the scenario of global warming. More detailed information, including predicted climate changes and consequences, is contained in the Annex. One thing is certain: no country will be immune from the impact of global warming, and in most cases the impact will be harmful. Those who focus on the theoretical benefits of global warming for particular countries are ignoring the net world-wide negative impact of climate change.

Chapter 2 summarizes the evolving technologies available to enable designers to create buildings which maximize comfort whilst minimizing their impact on the environment. Many different techniques and products are described, and the bibliography provided at the end of the book indicates sources of further information.

In Chapters 3 and 4, case studies are used to show how energy-efficient design techniques have been applied in practice. The range of studies encompasses domestic, commercial and institutional buildings. All have been completed during the current, final decade of the millennium, and serve as exemplars of the new tradition which must be fostered in architectural design.

The first three sections in the Recommendations section (pp98-109) deal with some of the issues involved in implementing change. They examine some regulatory policies, as well as issues which have been raised by post-completion and post-occupancy surveys, and offer some precautionary advice. It is important that designers benefit from feedback from buildings in which the architects and their consultants have set out to achieve an environmentally advanced package.

Biomorphic design represents a dramatic shift in design philosophy, so its development and optimization are still evolving. It is vital that the cause of biomorphic architecture is not undermined by a backlash due to a mismatch between aspiration and achievement.

We present a view of the future in the closing sections (pp 110-15). These sections include suggestions as to how designers will be challenged by the evolving consequences of global warming. They point to advances in technology which should make the autonomous building the rule rather than the exception, and they discuss the possible nature of the regulatory regime which might ensue following the accumulation of hard evidence for climate change.

Beyond all reasonable doubt

Of all the factors that threaten to undermine the sustainability of the planet, by far the most important is the accumulation in the atmosphere of so-called 'greenhouse gases'. This is causing the planet to warm, and this in turn is producing climate changes which may well be irreversible. These climate changes are already occurring at a rate which exceeds the adaptive capacity of some of the earth's biosystems. Ecological balance is often precariously poised, and small changes can have catastrophic effects.

The most important greenhouse gas is carbon dioxide, mainly because of the quantities which are being discharged into the atmosphere due to the burning of fossil fuels. The most authoritative evidence supporting the global warming/climate change scenario comes from the UN Inter-governmental Panel on Climate Change (IPCC) Scientific Committee. In its 1992 report, the committee stated:

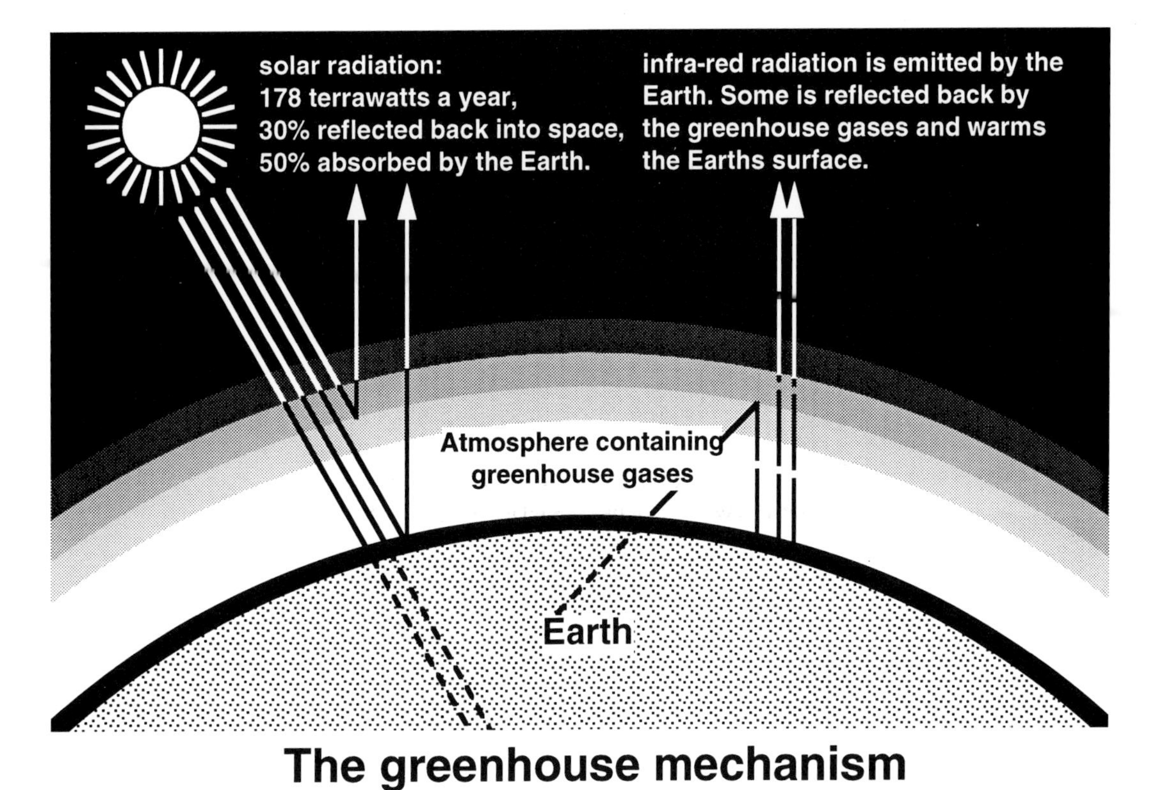

Figure 1.1
The greenhouse mechanism

'We are certain of the following: emissions [of carbon dioxide] resulting from human activity will result, on average, in a warming of the Earth's surface.' (Houghton et al., 1992)

In its 1995 report, the committee went further:

'Carbon dioxide remains the most important contributor to anthropogenic forcing of climate change; projections of future global mean temperature change and sea level rise confirm the potential for human activities to alter the Earth's climate to an extent unprecedented in human history; and the long time scales governing both the accumulation of greenhouse gases in the atmosphere and the response of the climate system to those accumulations mean that many important aspects of climate change are effectively irreversible.' (Meteorological Office, 1995, p.3).

The report concludes: 'there is a discernible human influence on global climate'.

The concept of the 'greenhouse effect' is frequently raised in discussions about climate change. However, it may be useful to offer a brief description of the mechanism involved.

The sun provides the energy which drives weather and climate. Of the solar radiation which reaches the earth, one third is reflected back into space; the remainder is absorbed by the land, biota, oceans, ice caps and the atmosphere. Some of the long-wave infra-red radiation reflected back from the earth is re-reflected to the earth by the greenhouse gases, and this is what causes warming of the planet. Figure 1.1 illustrates the process.

Under natural conditions, the solar energy absorbed by natural features is balanced by the energy re-radiated from the earth and atmosphere. Without this greenhouse canopy, the earth would be 33°C cooler. Since the Industrial Revolution, the combustion of fossil fuels and deforestation have resulted in an increase of 26 per cent in carbon dioxide concentrations in the atmosphere. The 1992 IPCC report predicts that a doubling of the concentration of carbon dioxide in the atmosphere compared to pre-industrial levels would lead to an average global temperature rise of as much as 4.5°C.

The first real evidence that there is likely to be a direct link between global temperature and atmospheric concentrations of carbon dioxide was revealed by ice core samples. Air trapped in ice dating back to 160,000 years ago provided a map of climate changes over this period. What is remarkable is the close correspondence between temperature and carbon dioxide concentrations. The graph which showed this correlation (see Figure 1.2) is based on one which appeared in *Nature* (Vol. 347, September 1990, p.139), and is now part of the iconography of climate scientists.

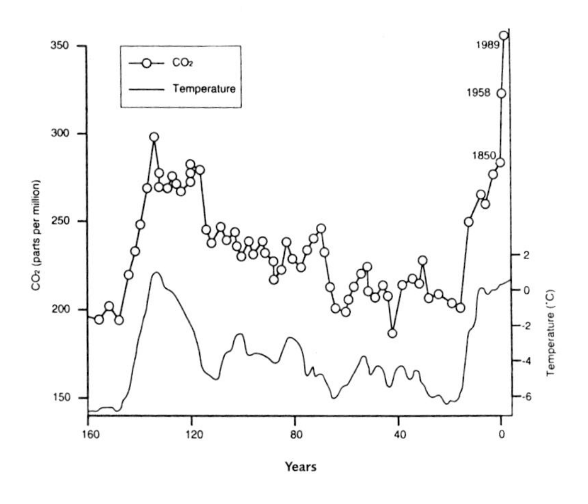

Figure 1.2
Correspondence between historic temperatures and carbon dioxide concentrations

The latest batch of ice core samples suggests that climate changes can occur gradually or catastrophically. During the interglacial periods, beginning 160,000 years ago, temperature rises produced severe climate perturbations, often over very short periods. From the point of view of building design, dramatic changes could well occur within the lifetime of a building.

As far as inhabitants of the United Kingdom are concerned, some compelling evidence for global warming is provided by the fact that, of the five warmest years since records began 337 years ago, three have occurred in the past ten years: 1989, 1991 and 1995.

Predictions regarding the impact of global warming are bound to be fuzzy. Nevertheless, the IPCC is confident about certain outcomes, the most critical being the rise in sea level due to the melting of land-based ice and the thermal expansion of sea water.

By 2100, the average rise could be as much as 110 cm. The worst-case scenario is that the ground-based West Antarctic Ice Sheet could be released due to warmer sea temperatures, resulting in a global sea level rise of 5 m. The consequences for small island states and some great maritime cities will be obvious. There will be massive migrations of people, coupled with the fact that much of the world's food-producing land is near coasts and will be lost at a time when world population is rising steeply.

There will be major shifts in rainfall patterns, resulting in forest death and increased desertification. Some of the most unpleasant pests and pathogens currently confined to subtropical regions will migrate to temperate zones. Higher temperatures will mean an increase in the scale and frequency of major storms, due to enhanced convective dynamics.

In 1993, the World Energy Council predicted that by 2020 there will have been a 44 per cent increase in greenhouse gases, largely due to the use of fossil fuels to power the economic rise of the developing countries (WEC, 1993). Soon fossil fuels will account for one third of global carbon dioxide production, and their output of this gas is set to double every 14 years. (A more detailed survey of the evidence for global warming and the likely climate changes is given in the Annex.)

Responsibility for carbon emissions is unevenly spread. To illustrate this, the annual per capita emissions of carbon for some key states are given in Table 1.1. Economic growth in China currently averages 9–10 per cent per year, and mostly relies on carbon-intensive energy.

Table 1.1 Annual emissions of carbon per capita in 1988

USA	5.85 tonnes
European Union	2.31 tonnes
Japan	2.35 tonnes
India	0.23 tonnes
China	0.65 tonnes

Source: Grubb (1990)

The only internationally agreed policy to curb the emission of carbon dioxide is the Rio Framework Convention, in which OECD countries agreed to stabilize their carbon dioxide emissions at 1990 levels by the year 2000, thus perpetuating the inequalities outlined above. By November 1996, only four countries were on course to meet the target: Sweden, Denmark, Germany and the United Kingdom. As far as the UK was concerned, this was more by accident than design, resulting from the shift from coal- to gas-fired electricity generation. The 1995 Berlin Convention sought to secure abatement commitments from the OECD countries extending beyond 2000. No agreement is yet in sight.

Where do designers fit into this rather depressing picture? Of the 6 billion tonnes of carbon currently emitted world-wide due to human activity, about 4.5 billion tonnes is attributable to the industrialized countries. Approximately half of this is due to buildings in one form or another. Architects and engineers have it in their power to reduce building-related carbon emissions by at least 60 per cent, which translates to 1.35 billion tonnes of carbon – at least equivalent to the savings envisaged by the Rio and Berlin conferences. At the same time, the built environment is the sector most amenable to environmental modification in the short to medium term. Where policymakers have failed, architects and engineers can succeed.

The task now is to examine in some detail the techniques for designing and constructing buildings that succeed in radically reducing carbon emissions.

The built environment

Are there any reasons why the information supporting climate change should be of special interest to the construction professions? There is no doubt that the mounting evidence of the changes that are occurring on a global scale due to the enhanced greenhouse effect should be a cause for general concern. However, this must be especially so for those with a professional interest in the quality of the environment. It is part of the nature of many who are concerned with the built environment to be optimists and idealists and to believe in the capacity of the physical environment to improve the quality of life. That quality is firmly coupled to the issue of sustainability, yet little is being done politically to halt accelerating global *un*sustainability.

Professions which claim to have a special concern for the environment must therefore be singularly challenged by the increasing amounts of hard evidence supporting the climate predictions of the UN IPCC Scientific Committee. At the same time, a catalogue of scientific evidence should help convince clients of the wisdom of commissioning environmentally friendly and energy-efficient buildings.

Whilst there may be considerable variation in responses on the moral plane, there can be no disputing the fact that the built environment is the main culprit in terms of depositing carbon into the atmosphere. As already stated, in developed countries, at least 50 per cent of all carbon emissions are attributable to buildings in one form or another. If you add the energy used in travelling to and from buildings, the figure is probably closer to 70 per cent.

There are two routes to the goal of reducing carbon emissions. First, there is the supply side of the system. The 'dash for gas' policy pursued by electricity generators in the UK in recent years has already achieved significant reductions in supply-side carbon emissions. For this and other reasons, it may be that the principle of free-market economics will not sanction too much interference in the wealth-generating capacity of the supply side. Therefore, the burden of responsibility for achieving radical reductions in carbon emissions will fall on the demand side, which means that buildings must be the principal target.

Where buildings are positioned, the way they are sited and their detailed design all have energy implications. Existing technology has enabled some architects to reduce energy demand in houses and commercial and institutional buildings by 60–70 per cent compared to current regulatory standards. A few housing associations and local authorities are doing valiant work in retrofitting older homes. Client demand may well seek to make such buildings the norm rather than the exception, and all construction professionals must be technically competent to meet this demand.

As will be described in Case Study 3.3, a house in Freiburg, Germany, has demonstrated that it is possible to avoid any input from energy services. New materials will transform the methods of design – for example, the use of aerogels (see page 35). In 1995, scientists in the USA announced a breakthrough in making aerogel production more commercially viable. At the same time, the cost and efficiency of photovoltaic (PV) cells are improving. Case Study 4.2 will describe one of the most ambitious PV retrofit undertakings: the façade PV treatment applied to the Northumberland Building, University of Northumbria, UK.

These are examples of the way technology is developing to make sustainable or biomorphic architecture an everyday possibility. Architects have always been adroit at exploiting new technologies to create new forms and spatial experiences. Environmental pressures could be said to be driving architecture into a new realm of opportunity.

At present, energy costs to consumers are relatively low because they benefit from a range of hidden subsidies. Delivered energy prices take no account of the external costs, such as the damage inflicted by acid rain. Harm to forests, crops, lakes, buildings and human health are all met from other sources. As global warming gathers pace, the climatic costs will rise steeply. When there are reliable methods for computing the cost of these externalities, it is expected that the price of fossil fuels will rise to compensate for both present and past damage. However, according to researchers at Cambridge University (Barker, 1993; Barker et al., 1994), even the most ambitious cost increase on this basis will not be enough to drive down consumer demand to a level approaching the carbon abatement target proposed by the IPCC. This will only be achieved by the added burden of a carbon tax. They suggest that well within the lifetime of buildings now on the drawing board, fossil energy costs will need to rise by a factor of seven in real terms if there is to be any hope of halting global warming. It therefore becomes a matter of professional responsibility to clients to explain that energy-efficient buildings are not only good for the planet, they also make good medium- to long-term economic sense. When you add to this the fact that low-energy-use, naturally-ventilated buildings, when properly designed, create better comfort conditions for work and home, then the arguments in favour of biomorphic design become irresistible.

On the construction side of the industry, much can be done to reduce the environmental impact of buildings. A significant number of building materials are capable of being recycled, and an emphasis on recycling should inform both the design and construction processes. At present, the transport industry is leading the way in this respect – for example, German manufacturers such as VW-Audi are allowing for materials recycling in the design of their cars.

It is estimated that 20 per cent of building costs result from waste. Again, both designers and contract managers should share the aim of eliminating waste.

As regards the selection of materials, information about the energy embodied in materials is becoming increasingly available, and this should be a decisive factor in the choice of materials, including the energy costs of transporting materials and components to the site. At the same time, it is necessary to be sparing in the use of non-renewable resources such as mined aggregates (see Anink et al., 1996).

All this raises questions for which there are still only partial answers. This book identifies some of the signposts on the route to a sustainable future, and illustrates some of the buildings which are leading the way.

2: Evolving solutions: Technologies and techniques

Chapter 1 clearly established the overwhelming need to address energy and environmental issues in building design, and presented examples of attitudes and developments which should be encouraged. We will now focus on providing specific information on a wide range of technologies, products and techniques which can be employed within and around buildings to create energy-efficient and environmentally aware design solutions.

No element of building design can be considered in isolation, though for convenience and ease of understanding, individual aspects of design are examined in this chapter, before progressing to case studies of buildings which, by their nature, encompass a more holistic approach in Chapters 3 and 4. It is important that those involved in the building design process take an active interest in these technologies in order to demonstrate and develop their use. Good designers ought to be able to accommodate these technologies within quality architectural design, and there is a need to provide sufficient explanation to allow and encourage exploitation of their potential.

Most of the energy used in non-industrial buildings is employed in conditioning the climate of the interior environment – for heating, mechanical ventilation and air conditioning (HVAC) , and also for artificial lighting. The aim must be to reduce the negative and maximize the positive impacts of the external climate, so that the requirements for conventional HVAC services can be minimized. Novel technologies, such as solar photovoltaic cells, and technologies which improve the operation of the climate control services are also discussed, but it should be remembered that the main constraints are set by fundamental building design decisions.

Overriding limitations on the technologies which it is possible to apply in any given situation are the climate, the building type and its function. These parameters normally fall outside the boundaries of the design process, and so will not be addressed here. However, the descriptions should permit application to any relevant situation, should the need arise.

Where possible and appropriate, checklists of the important features for consideration under each heading are provided, to enable the reader to access the significant details quickly and efficiently.

Site design and layout

Since the site constrains what can be achieved in terms of energy-efficient and environmentally sound design, it is important to give careful consideration to site planning in order to maximize its potential. There are several benefits to integrating climate and site development, such as reductions in both winter heating costs and the occurrence of summer overheating which otherwise leads to increased cooling requirements. A further benefit is the general increase in occupant comfort. In addition, the durability of building materials may be increased, and a more pleasant environment in the vicinity of the building may be created.

Building designers can have little specific influence on the overall (or 'macro') climatic situation, except in the senses laid out in Chapter 1. They may also be able to have only a limited effect on the features within the locality which will affect the climate on a medium regional (or 'meso') scale. Factors to consider here would include the nature of the surroundings (city-centre; suburban or rural), and topographical features such as hills and valleys which affect exposure to wind. However, site design decisions do have an important influence at the micro-climatic scale, and need to be understood and appreciated.

Factors to consider include:

- the effects of, and relationship to, buildings and other influences at the site boundary;
- the positioning of access roads and pedestrian pathways;
- planting of trees and other vegetation;
- the positioning of walls, fences and other obstructions;
- the orientation of building plans and façades in relation to the sun;
- massing of buildings and building grouping;
- spacing between buildings;
- sunlight and shade;
- the wind environment;
- maximizing external views
- the positioning of main entrances – this is important because the local ambient environment around such entrances should act as a transition between inside and outside conditions;
- landscaping, planting and installing pools to enhance natural cooling;
- environmental noise and pollution;
- security and health issues related to building openings.

Most of these factors are discussed in more detail in later sections. However, the last two (which do not fit neatly into other categories) are considered briefly below.

Building openings may be utilized to permit air and heat transfer, but they also permit the intrusion of external noise. Their positioning must therefore take account of any sources of environmental noise, such as nearby roads or noise from nearby buildings and processes. City-centre locations, and even some suburban site locations, also pose air pollution difficulties. Such pollution is difficult to avoid, and is likely to result in the design of a well-sealed building with less flexibility in environmental design. In some urban areas, it may be very difficult to exploit natural ventilation in the detailed building design because of air pollution, which is usually associated with noise (see Case Study 4.10). These problems can be alleviated by adopting an appropriate basic layout.

Openings also present weak points in the building envelope from a security point of view – site design needs to avoid the problems which arise with façades which cannot be secured or monitored easily. The location of sensitive buildings within the site needs to be considered, and landscape/vegetation can help or hinder visibility and security.

Planting of suitable vegetation to repel insects such as mosquitoes can be used in environments where they may pose a hazard – in particular to reduce their presence around open windows and ventilation openings.

The remaining aspects of site design have particular relationships to solar heat gain and ventilation air flows, and are dealt with as they arise in the appropriate sections below.

Climate-sensitive design

The phrase 'climate-sensitive design' has attained a level of credibility and acceptability in recent years, to the extent that major building design practices pay more than a passing interest in what can be exploited. Such design is sometimes referred to as 'solar architecture', 'climatic architecture' or 'bioclimatic architecture' (though perhaps more correctly it should be called 'biomorphic architecture'). In essence, it is the acknowledgement that the interaction of solar and other climatic factors with the building envelope will determine the basic internal environmental conditions and thus the extent to which HVAC services are needed and, ultimately, the comfort level of the occupants.

Clearly, it is possible to overcome many of the negative effects of a poor design which is insensitive to the climate by the use of heating, ventilation and, worst of all, air conditioning. Although there are circumstances where these systems are unavoidable, the extravagant use of them is irresponsible, and should be unacceptable in the present day.

Design features

The adoption of climate-sensitive design recognizes the important impact of the following factors in building construction:

- designing an appropriate building form which does not lead to unnecessary overshading of one building by another;
- utilizing suitable construction techniques such as the relative positions and thicknesses of insulation materials to maximize beneficial heat gains or to exclude excess heat;
- adapting the internal layout to the climate and building orientation so that rooms or spaces with specific functions are located adjacent to the most appropriate façades;
- dividing buildings into thermal zones with buffer areas such as balconies, verandas, atria, courtyards and arcades – though these divisions should avoid creating barriers to cross-flow ventilation where this is required;
- choosing and positioning appropriate building materials within the internal and external fabric, particularly where thermal mass effects can be used to dampen temperature fluctuations;
- selecting the location, size and type of openings in the building envelope (including glazing type) to exploit advantageous solar gain e.g. glazing facing south in northern latitudes is easier to shade;
- integrating the building design with the site design, so that site obstructions and tree or noise problems are not exacerbated;
- integrating with the occupants' needs and expectations – generally, occupants are prepared to accept less than ideal levels of comfort providing they have some degree of control over the environmental design, and understand it.

Since the potential of many of these features can be exploited within the normal building process at minimal additional cost, there is no reason why such design techniques should not be adopted – indeed, there are numerous examples of the application of climate-sensitive design in contemporary architecture.

Materials, thermal mass and energy storage

One of the key elements of good climate-sensitive design involves the choice of appropriate materials for energy absorption. In many climates, ambient temperature varies in a daily pattern, with maxima during the middle of the afternoon and minima overnight or in the early morning. Maximum solar input also occurs during the middle of the day. By choosing suitable building materials with appropriate 'thermal mass' (the accepted term for the overall effect of the amount of material combined with its thermal capacity), heat can be stored and temperature fluctuations reduced, resulting in more acceptable thermal conditions.

Recent innovations include the promotion of cold storage, whereby building materials are pre-cooled overnight so they are able to absorb and counteract some of the following day's overheating. A Scandinavian system known as Termodeck consists of an interconnected network of tubes running through floor slabs, through which cool air can be circulated at night to pre-cool them.

Water, which has a high thermal capacity, may also be used to absorb heat. One of the more exciting possibilities involves the use of phase-change materials. When water changes from a solid (ice) to its liquid form, it absorbs a large amount of heat before its temperature increases. Phase-change materials proposed for use in buildings work on a similar principle. However, the temperature at which they change state is close to that which is acceptable for thermal comfort (20–30°C). In this way, substantial amounts of heat can be absorbed to counteract temperature rise and discomfort. Of course, the reverse cycle also occurs – the heat is liberated from the phase-change materials when they return to the solid state on cooling.

Experimental work is now being carried out to develop suitable materials which can safely be incorporated into building components. One option is to impregnate plasterboard with such phase-change material, where it is held in place even when in its semi-liquid state.

Integrating climate-sensitive design

Designers should not shy away from the exploitation of climate-sensitive design. What is sometimes forgotten is the extent to which such design principles were utilized in the traditional, vernacular architectural styles of many areas of the world. A number of scientific studies have confirmed that traditional designs were often extremely well suited to the rigours of their local climate. In many countries, that association between climate and design was degraded by the need for rapid construction techniques (partly caused by urbanization and population migration) and the ready availability of fuels and building services systems to overcome any detrimental effects. The environmental costs and, increasingly, the monetary costs of such short-sighted design philosophies are now being recognized and alternatives sought. Reluctance to re-employ traditional techniques may arise from stylistic concerns and from the need to be able to predict with some certainty the outcome in comfort terms of using less comprehensively serviced buildings. Both of these concerns can and should be overcome.

It has been traditional to define climate-sensitive design as 'passive solar' design. This reflects the pre-eminent role of solar heat gain in such an approach, but it implies a rather limited set of solutions, especially since the prime building type has tended to be dwellings. In the following explanations, a number of integrated solutions are discussed, including passive solar design.

Solar design

Whether it is important to encourage or prevent solar radiation impinging on the building, it is necessary to appreciate the degree to which solar access is available, so that the likelihood of solar heat gain can be determined. At the earliest stage of design, one must consider the following parameters in relation to the site:

- the sun's position in relation to the principal façades of the building (solar altitude and azimuth);
- site orientation and slope;
- existing obstructions on the site;
- the potential for overshadowing from obstructions outside the site boundary.

For the development itself, the following factors need to be considered:

- grouping and orientation of buildings;
- road layout and services distribution;
- proposed glazing types and areas, and façade design;
- the nature of internal spaces into which solar radiation will penetrate.

One method to evaluate solar access involves the use of some form of sun chart. The type most often used is the stereographic sun chart (see Figure 2.1), in which a series of radiating lines and concentric circles allow the position of nearby obstructions to insolation, such as other buildings, to be plotted. On the same chart, a series of sun path trajectories is also drawn (usually one arc for the 21st day of each month), and the times of the day are also marked. The intersection of the obstructions' outlines and the solar trajectories indicate times of transition between sunlight and shade. Normally, a different chart is constructed for use at different latitudes at about 2° intervals.

Sunlight and shade patterns cast by the proposed building itself should also be considered. Graphical and computer prediction techniques may be employed, as

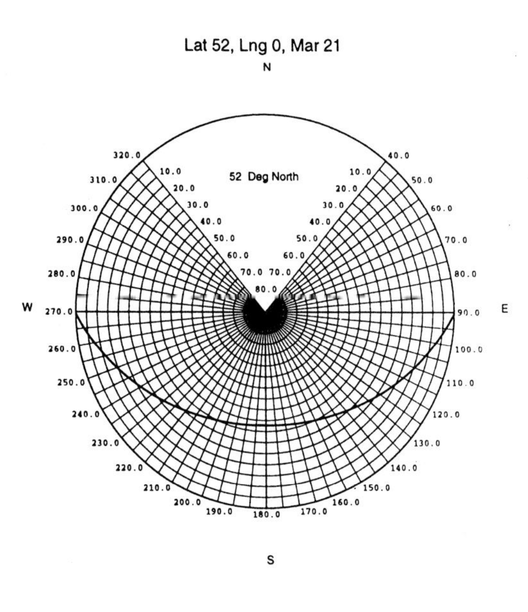

Figure 2.1
Example of stereographic sun chart, showing sunpath line for 21 March

well as techniques such as testing physical models with a heliodon.

The spacing between buildings is very significant if overshading is to be avoided during winter months, when solar heat gain is most advantageous. The following factors should be considered and acted upon.

- For parallel rows of dwellings, the spacing between the rows has a greater impact on shading than the orientation of the rows.
- Particular attention must be paid to combinations of spacing and slope on the site.
- If overshading is to be avoided in winter for buildings at a latitude of 50°N, rows on a 10° north-facing slope must be more than twice as far apart as the same buildings on a 10° south-facing slope.
- In northern latitudes, south-facing slopes are obviously preferable.
- Benefits may be obtained by placing individual, or low-density, buildings to the south side of a site, and tall buildings, rows or terraces towards the north.
- For housing schemes, rows of terraced dwellings are better placed on roads with an east–west axis, whilst detached housing is better arranged on a north–south axis.

Trees can also provide obstructions to solar access, but if they are deciduous, they can perform the dual function of permitting insolation on the building during the winter whilst providing a degree of shading in the summer. Again, the spacing between the trees and the building is critical.

The building techniques which exploit solar heat gain are normally defined as 'passive solar design'. This is further subdivided into three rather broad categories:

- direct gain;
- indirect gain;
- attached sunspace (or isolated gain).

Sunspaces are subdivided according to whether their main benefit is of a direct or isolated nature. In this book, the term 'attached sunspace' is preferred because of its significance in promoting an aesthetically pleasing design option.

Each of the three categories relies in a different way on the 'greenhouse effect' as a means of absorbing and retaining heat. The greenhouse effect in buildings mimics global environmental warming on a small scale. In buildings, the incident solar radiation is transmitted through façade glazing to the interior, where it falls on the internal surfaces and is absorbed by them, warming them up. However, re-radiation of heat back through the glazing is inhibited by the fact that it has a much longer wavelength than the solar radiation because it is emitted from surfaces at a much lower temperature, so the glazing reflects it back into the interior.

The three categories are dealt with in more detail below, but they mainly apply to domestic buildings. Commercial and institutional buildings, which operate to different occupancy and heat gain schedules, cannot be analysed in the same way; indeed, provision of cooling may be a more important requirement than heating, even in relatively cool climates. Cooling and ventilation appropriate to such buildings are analysed later in this chapter.

Direct gain

In the direct gain design technique, it is desirable to concentrate the majority of the building's glazing on the sun-facing façade (the south façade in northern latitudes). Solar radiation is admitted *directly* into the space concerned. These are the main design features.

- Apertures through which sunlight is admitted should be on the solar side of the building, within about ± 30° of south for the northern hemisphere.
- Windows should avoid facing west, as this leads to the risk of summer overheating.
- Windows should be double- or triple-glazed with low-E glass.
- The main occupied living spaces should be located along or behind the glazed façade.
- The floor should be of a high thermal mass to absorb the heat, and to provide thermal inertia which reduces temperature fluctuations inside the building.
- For normal daily cycling of heat absorption and emission, only the first 100 mm or so of thickness is involved in the storage process; thicknesses greater than this provide only marginal improvements in performance.
- If acceptable, the floor should be dark in colour (to aid heat absorption).
- Insulation should be below, not above the floor.
- A vapour barrier should always be installed on the warm side of insulation.
- Thick carpets should be avoided over the main sunlit and heat-absorbing portion of the floor.

During the day and into the evening, the warmed floor should slowly release its heat, and the period over which this occurs is ideal for domestic circumstances, when the main demand for heat is in the early evening. Figure 2.2 illustrates the main design features.

If carpeting is widespread, it acts as an insulator which prevents the heat being absorbed by the floor, and this leads to more immediate and unwelcome overheating. However, the floor mass should be insulated beneath and at the sides using conventional insulation materials.

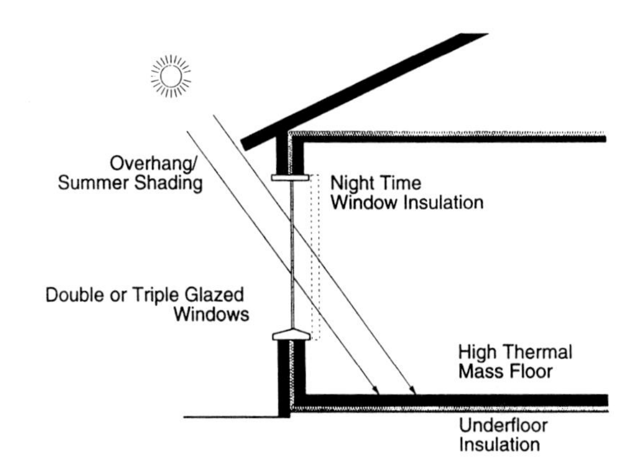

Figure 2.2
Passive solar, direct gain

As far as the glazing is concerned, the following features are recommended:

- The use of external shutters and/or internal insulating panels might be considered to reduce night-time heat loss.
- To reduce the potential for overheating in the summer, shading devices should be available as external overhangs or awnings, as louvres or as internal blinds (external shading is more effective than internal, since heat absorbed by the shading device is more easily dissipated).
- Heat-reflecting or absorbing glass may be considered to reduce overheating, though this also reduces heat gain at times when it may be beneficial.
- Diffusing glazing elements may be employed if a more even spread of heat gain to the internal surfaces is required, but many of the aesthetic benefits of the view outside will be lost.
- Light shelves can help reduce summer overheating while improving daylight distribution.

Direct gain is also possible through the glazing located between the building interior and the attached sunspace or conservatory; it also takes place through upper-level windows of clerestory designs. In each of these cases, some care needs to be exercised regarding the nature and position of the absorbent surfaces.

In the UK's climate and latitude, these general rules of thumb apply:

- The room depth should not be more than 2.5 times the window head height.
- The glazing area should be about 25–35 per cent of the floor area.

Indirect gain

This form of design places some building element between the incident solar radiation and the space to be heated, so that the heat is transferred in an indirect way. There are many forms and variations of passive design within this classification, and it is not possible to describe them all here. The main options usually employ a thermal storage element – often a wall placed behind glazing facing towards the sun – and this controls the flow of heat into the building. The main factors in the function of the design are:

- A high-thermal mass element is positioned between the sun and internal spaces; the heat absorbed conducts slowly across the wall, and is liberated to the interior some time later.
- The materials and thickness of the wall are chosen to modify the heat flow (for domestic situations, the flow can be delayed so that it arrives in the evening, matched to occupancy periods); typical thicknesses are 20–30 cm.
- Glazing on the outer side of the mass wall is used to provide some insulation against heat loss and help retain the solar gain (making use of the greenhouse effect).
- The area of the thermal storage wall element should be about 15–20 per cent of the floor area of the space into which it emits heat.
- In order to derive more immediate heat benefit, air can be circulated from the building through the air gap between the wall and glazing and back into the room; in this modified form, this element is usually referred to as a 'Trombe wall'.

Figure 2.3
Passive solar,
indirect gain – Trombe wall

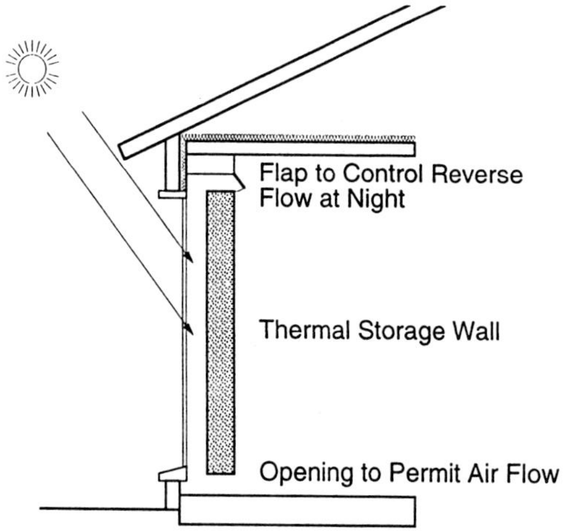

In countries such as the UK which receive inconsistent levels of solar radiation throughout the day because of climatic factors, the option of circulating air may be of greater benefit than awaiting its arrival after passage through the thermal storage wall. Figure 2.3 shows a sketch of a Trombe wall. The air circulation must be restricted at times of low solar radiation and cold temperature, such as overnight, since this can induce reverse flow and heat loss.

At times of excess heat gain, the system can provide alternative benefits: the circulating air can be vented directly to the exterior, carrying away its heat while drawing outside air into the building from cooler external spaces. The overheating risk can be reduced by applying some of the options discussed above in relation to direct gain systems.

Alternatives to the conventional indirect gain devices involve the substitution of alternative materials for the heat-absorption media, and the modification of the position of the absorber. In a number of cases, the heavy masonry material of the wall is replaced by water. Water has a higher heat capacity per unit volume, and is thus able to absorb larger amounts of heat. There are some difficulties, including the obvious ones such as the risk of leaks or punctures to the water containers. From a thermal point of view, buoyancy currents occurring in the water can reduce the effectiveness by permitting increases in convection heat transfer within the water container. There are times when this phenomenon might be advantageous, but it must be controlled.

The use of water in the roof has also been proposed for lower-latitude countries – a 'roof pond' system. At such latitudes, the horizontal roof provides a better absorber than vertical walls, and in addition, it may be possible to exploit night-time cooling of the water mass during the summer months. This can provide a means of reducing overheating the following day. The dynamic use of insulation panels or covers is an important aspect of the successful functioning of this technology. During the winter, the water mass is uncovered during the day, but covered at night to retain heat. For summer months, the opposite schedule is used: the water is covered during the day so it absorbs heat from the building, and and it is uncovered at night to allow heat loss.

Indirect gain options are often viewed as being the least aesthetically pleasing of the passive solar options, partly because of the restrictions on position and view out from remaining windows, and partly as a result of the dark surface finishes necessary for the absorbent surfaces. As a result, this category of the three prime solar design technologies is not as widely used as its efficiency and effectiveness warrant.

Attached sunspace/isolated gain

The attached sunspace often takes the form of a conservatory or greenhouse. Variations in design promote a range of functions: extra living area, solar heat store, ventilation pre-heater, and buffer space or entrance vestibule. The conservatory seems widely favoured in architectural design as a visually pleasing and thus desirable passive solar option. Figure 2.4 illustrates the basic components of the design.

Despite their popularity with both designers and the public, the environment within sunspaces is very difficult to control, and it is more difficult to demonstrate reduced energy consumption and increased comfort. In the UK, for instance, the main attraction of conservatories is the increased living space which they provide. Little attention is paid to their orientation, and it is common to find conventional heating systems installed within them to compensate for poor design or to override bad weather conditions.

In a study of a range of architects' design proposals for passive dwellings in the UK, increased energy use was predicted in several cases due to the need to heat large conservatories which had been designed as an integral part of the living accommodation (ETSU, 1993).

In order to obtain the optimum in performance, the conservatory may need to be capable of isolation from the rest of the building. This technique adds a degree of controllability, and can reduce the problems of heat loss in winter and heat gain in summer. The area of glazing embodied in the sunspace should be 20–30 per cent of the occupied floor area to which it is attached (see Case Study 3.4).

Air flow paths to and from the conservatory, and between it and the building itself, must each be well designed and controlled to obtain the best performance. This is also true of the general class of isolated-gain designs which commonly rely on heat transfer using air as the medium. The moisture content of the air within such designs needs to be considered if the problem of condensation on the coldest surfaces within the design (often the conservatory roof) is to be avoided.

Despite the problems, it is likely that the attached sunspace passive solar design option will continue to be popular, in which case much greater attention needs to be paid to its energy aspects.

Figure 2.4
Passive solar – attached sunspace

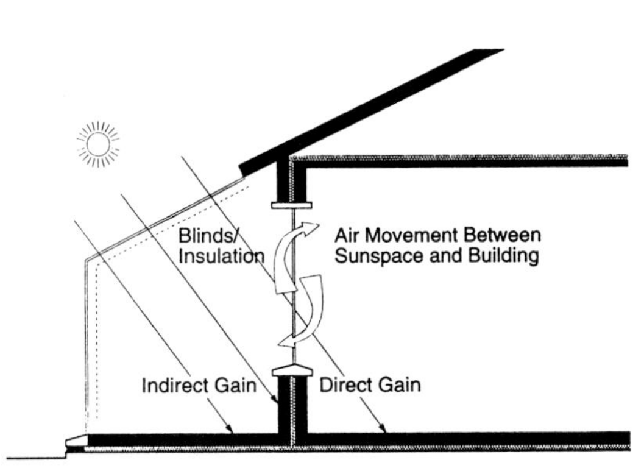

Assessment of passive solar design

In order for building designers to have confidence in the attractions and benefits of passive solar design, some form of assessment may be required to demonstrate these in a quantitative manner. Two such methods are the Solar Load Ratio (SLR) Method (or one of its derivatives), devised in the USA, and the New Method 5000, which was conceived in Europe. Given basic data, each will allow a month-by-month analysis to be performed, which can then provide an estimate of annual energy savings. The methods are not too onerous, yet they still tend to be ignored by architects, either because of lack of skill or shortage of time.

Whilst the three main categories of passive solar design, along with their subdivisions, are most usually applied within domestic-scale designs, similar principles have been analysed with reference to commercial developments. One assessment method which addresses this sector is the Lighting and Thermal Value of Glazing Method (the LT Method – Baker and Steemers, 1994), developed in the UK. This method reduces a building to an orthogonal plan with core and perimeter zones. The perimeter zone is that which is subject to significant external climatic influences on its lighting, heating and cooling requirements. Perimeter zones are classified by orientation and depth, and are defined as passive zones. The LT Method, which has so far been developed for use in the European climate, permits a straightforward prediction of likely energy use for lighting, heating and (if specified) cooling services on an annual basis.

Whilst such an approach is somewhat simplistic, it does provide a quick guide to energy consumption by indicating optimum window size and orientation at the initial design stage. It is therefore valuable in determining the basic plan form.

Alternative assessment techniques can be used, but are more complex, requiring a higher level of input and more time. Graphical/calculation techniques are available, though the attitude of many designers seems to be that if more sophisticated analysis is required, then some form of computer simulation of the building's energy flows and performance should be carried out. This means that the process requires specialist help, which is not conducive to enabling the architect to arrive at the optimum solution quickly. Improvements in 'user-friendliness' and simplification of data input are occurring, though extreme care is needed to ensure input assumptions and outputs are understood.

For complex analysis, a number of program suites now exist, though some critics question the absolute correctness of the results, and hence the reliance which can be placed on them. The simulation programs require time-consuming input procedures and interpretation of the data output, but can provide a sophisticated and seductive analysis of design proposals. They have great value as comparative tools and for use in complex building situations, and they are constantly evolving, yet one should always be aware of their limitations.

The main problem facing architects and building designers in assessing passive design lies in the need to use consultants or to seek specialist advice. Qualitative analyses may indicate comparative benefits, but quantitative analyses must be made much more accessible; this accessibility must also be accompanied by increased robustness and accuracy in the assessment results.

Biomorphic design

One analysis technique which has been developed over a number of years uses a Building Bioclimatic Chart. This is based upon a conventional psychrometric chart, onto which the boundaries of the thermal comfort zone, expressed by air temperature and moisture content (humidity), are plotted. Variations in external climatic conditions can be plotted on the same chart, normally in the form of a range of daily conditions experienced during each month of the year. In an ideal climate, the prevailing climate plots would all fall within the comfort zone, thus obviating the need for heating or cooling. However, this is not the case for the vast majority of real-life situations. A further technique can be applied which allows consideration of the effects of a number of building design options which could mitigate the effects of external climate on internal conditions. The main options include:

- using increased thermal insulation levels in the building fabric;
- increasing the effective thermal mass by the choice and location of construction elements and materials;
- encouraging increased natural ventilation and associated heat transfer through building openings;
- increasing night-time ventilation when external

temperatures are likely to be at their lowest to promote cooling;
- using evaporative cooling on building surfaces or at openings;
- adapting the design to promote radiative cooling from building surfaces.

The value of employing such design techniques is illustrated graphically by the modification of the boundaries of the comfort zone to show the range of external conditions over which it is possible to create internal comfort using specified design techniques. An example of how the techniques can be represented on a chart is shown in Figure 2.5. The chart enables the most useful building strategies for a particular climate to be identified quickly. This method has great visual appeal, and whilst there may be some disagreement over the exact boundary calculations, it has become widely used as a powerful tool for climate-sensitive design.

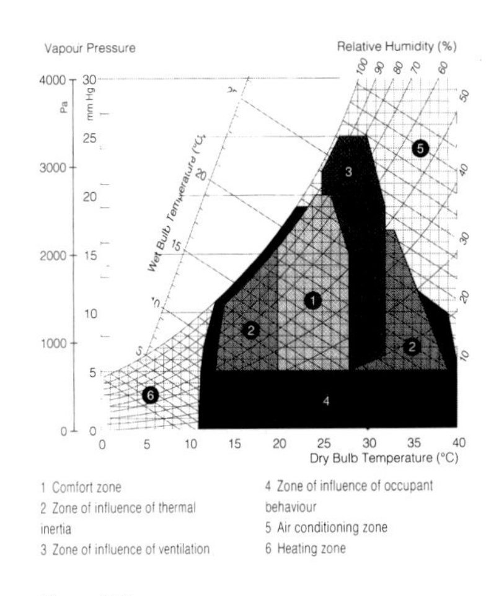

Figure 2.5
Building Bioclimatic Chart

Insulation

Heat flow through building components can be modified by the choice of material. The main heat transfer process for solid, opaque building elements is conduction. Thermal insulation is used to reduce the magnitude of heat flow in a 'resistive' manner. Since air provides good resistance to heat flow, many insulation products are based upon materials which have numerous layers or pockets of air trapped within them. Such materials are thus low in density and light in weight, and in most cases are not capable of providing structural support for the building. Generally, the higher the density of the material, the greater the heat flow. Since structural components are often, of necessity, rather high in density, they are unable to provide the same level of resistive insulation. Care should be exercised in providing additional layers of insulation around them to prevent such elements acting as weak links (often called 'cold bridges') in the thermal design.

Increased levels of insulation are a very cost-effective way of reducing energy consumption for heating. In several domestic and other small buildings, it has already been demonstrated that the additional costs of insulation can be offset against much reduced costs for the heating system – for example, two or three small unit heaters could be sufficient, rather than employing a whole-building central boiler and radiator system.

When specifying insulation materials, it is important to avoid those with detrimental environmental impacts, such as those which employ CFCs (chlorofluorocarbons) in their production process. There are three broad classifications of insulation material:

- *natural organic* – mainly processed vegetation-based materials, which must be treated to avoid rot and vermin attack;
- *inorganic* – these include common products based on silicon and calcium (glass and rocks), often in the form of fibrous matting;
- *synthetic organic* – as the name implies, these can be synthesized from organic feedstocks based on polymers; a wide range of compounds are available to meet particular requirements, with varying properties and costs.

Thermal heat transmission potential through a building is usually referred to by its U value: the lower the U value, the better the insulating performance. It takes into account thermal resistances attributable to building materials, air cavities, and internal and external surfaces.

Superinsulation

In recent years, attention has focused on the use of very thick layers of insulation within the building fabric in order to minimize heat flow. This technique has become known as 'superinsulation'. The use of superinsulation has so far been most fully demonstrated at a domestic scale, in houses and other small buildings. One reason for this may be the problems of overheating experienced in many larger, deeper-plan commercial buildings – problems which override the benefits of reduced winter heating requirements. In the future, however, buildings which exhibit less tendency to overheat due to better environmental design may modify the priorities and make superinsulation attractive in all circumstances where buildings experience cold seasons.

Superinsulation is associated with several design features:

- U values are less than 0.2 W/m²°C for all major non-transparent elements;
- U values are often below 0.1 W/m²°C.
- limits on the thicknesses which can be used are sometimes set by accepted construction techniques – for instance, by allowable cavity widths in cavity wall construction;
- a flexible definition of 'superinsulation' allows the specification of a maximum overall building heat loss which permits 'trade-offs' within certain limits, rather than individual component values;
- in low-energy housing, typical thicknesses of insulation material are about 150 mm in walls and 300 mm in roofs, whereas superinsulated walls may have 200–250 mm of insulation;
- if considering superinsulation, the other major source of heat transfer – air movement – should not be ignored.

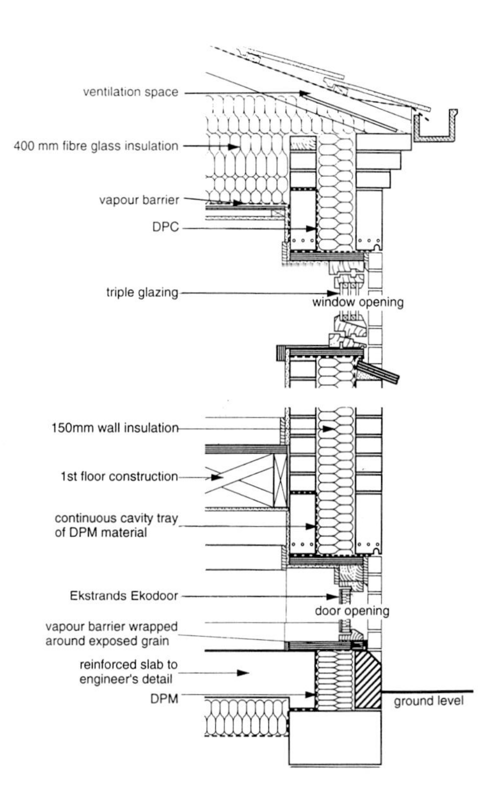

Figure 2.6
Partial cross-section of a superinsulated building (after Brenda and Robert Vale)

Several of the case studies used in this book are classic examples of the application of superinsulation principles. Figure 2.6 shows some construction details of a superinsulated building.

Transparent insulation materials

Transparent insulation materials (TIMs) enhance solar heat gain while reducing heat loss by conduction and radiation. The technology has similarities to the passive solar thermal mass wall designs described above, except the gap between the glazed outer cover and the surface of the wall which faces into it contains transparent insulation, rather than just air. The insulation allows transmission of the incoming solar radiation, but acts as a barrier to conductive and radiative heat loss, retaining absorbed heat very effectively. Studies have indicated that a net reduction in heat loss from south-facing walls of about 75 per cent can be achieved.

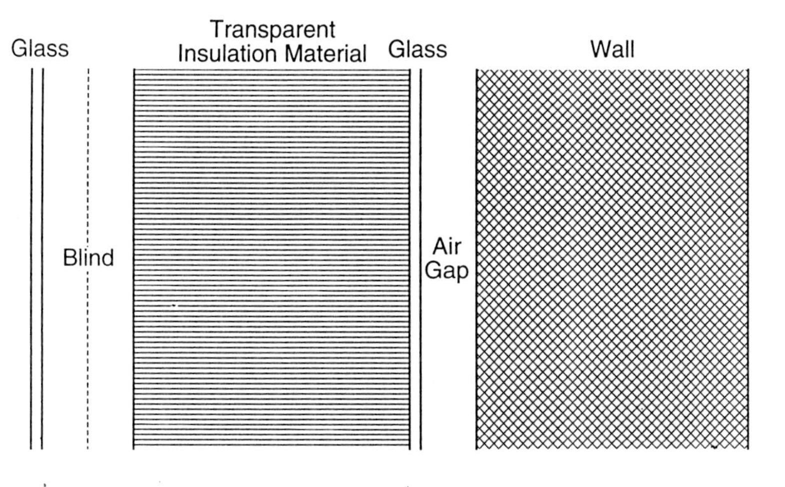

Figure 2.7
Cross-section of transparent insulation material

A number of different materials are employed for the insulation, and minor variations of position are possible. A design example is shown in Figure 2.7. Whilst the component includes insulation, it can still be a source of heat loss, though most of the difficulties with the design seem to relate to the control of heat gain when it is not required. Shading elements may be lowered into position, though these have been known to fail.

In principle, TIMs are an excellent idea, and development of thermochromic glass should eliminate the need for mechanical blinds. This is a laminated form of glass containing a chemical which turns translucent at 30°C, thereby excluding 70 per cent of solar radiation.

Aerogels encased withing a glazing sandwich might also be considered as a form of transparent insulation material. They are discussed in greater detail on page 35.

Insulation: Technical risks

The use of high levels of insulation brings with it some risks. Some problems relate to the presence within the building construction of moisture which, because the temperature gradient has been changed by the presence of insulation, condenses to form water. This can lead to several difficulties, such as rotting, rusting or other degradation of components, and in addition, can pose a safety risk if it comes into contact with electrical circuitry. Some insulation materials absorb moisture, and when wet, their insulating effect is very much reduced.

If substantial variations exist between the insulation levels of different parts of the building fabric, this creates weak links which then become the main heat loss flow paths. In particular circumstances of high heat transfer, they are referred to as 'thermal bridges' or 'cold bridges', and condensation is most likely to occur on the inner surfaces of such cold bridges. Care must be taken to ensure that insulation either overlaps or is continuous. The main focuses for concern tend to be at the joints between different components, for example:

- the joint between roof and wall;
- the joint between wall and floor;
- around windows and doors, particularly frames and lintels;
- around apertures for building services – electrical, water, drainage, etc.;
- at positions where structural elements connect with roofs, walls and floors.

When considering floors, the majority of the heat loss occurs at their exposed edges. Particular attention must therefore be paid to adequate and well-designed insulation details at floor edges.

The use of vapour control layers becomes more important as insulation levels rise, since the appropriate construction and positioning of such layers reduces the risk of condensation. It is advisable to carry out a technical assessment of condensation risk if this is suspected to be a problem, but perhaps it is more important to design components correctly and ensure that the construction is performed as specified. A large fraction of the reported faults associated with condensation are attributable to poor workmanship. As stated earlier, as a general principle, vapour barriers should always be on the warm side of the insulation, otherwise they will be a source of condensation.

Active solar thermal systems

A distinction must be drawn between passive means of utilizing the thermal heat of the sun, discussed earlier in this chapter, and those of a more 'active' nature. These active systems are able to deliver higher temperatures and a more controllable energy supply. However, a penalty is incurred since energy is required to control and operate the system – the 'parasitic energy requirement'. A further distinction is the difference between systems using the thermal heat of the sun, and systems, such as photovoltaic cells, which convert solar energy into electrical power.

In the past, active technologies have often been added on to the building after design and construction has been completed, which usually results in a less pleasing aesthetic product. There is some evidence to suggest that a more integrated design approach is beginning to gain ground, however, and this is to be encouraged.

The most common type of active solar thermal technology is the flat plate collector. Construction details vary from manufacturer to manufacturer, and depend on whether air or, more usually, water is being circulated through the collector to extract the heat absorbed. Figure 2.8 shows a typical cross-section. Four main components are common to most designs:

- a transparent cover plate;
- an absorber plate;
- an insulated enclosure box;
- flow passages/pipes to permit and control the flow of heat-extraction media.

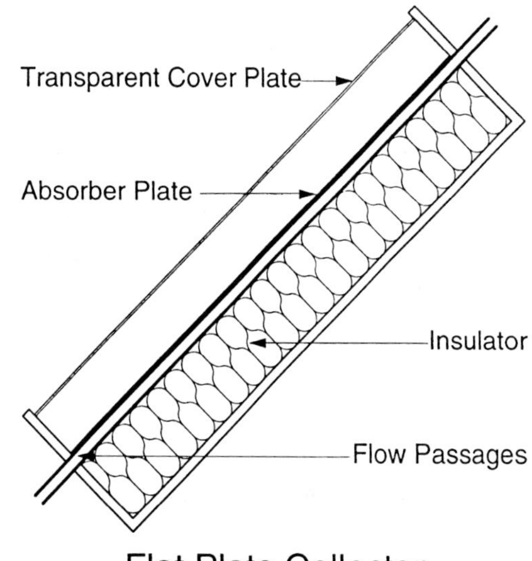

Figure 2.8
Cross-section of flat plate solar collector

There is often a trade-off in collector design between sophisticated, high-efficiency designs, and lower-efficiency but cheaper options. Limitations on use may also be set by the availability of suitable mounting surfaces of appropriate size, orientation and inclination. The following are generally accepted guidelines:

- Orientation should be close to due south in the northern hemisphere.
- Plates tending to the horizontal are most suited for summer heat provision.
- Plates tending towards the vertical are better for winter collection.
- An angle of inclination approximating to the latitude of the site gives reasonable year-round performance.

The principal function of many flat plate collector systems is to collect and deliver sufficient heat for domestic hot water, mainly during the summer months. The sheer physical area of collector required for space heating applications inhibits flat plate collector use in many situations. Furthermore, in order to guarantee heat during all months of the year, a substantial heat store is required to retain excess heat from summertime for winter use. For a three- or four-bedroom house in the UK, the collector area would need to cover all available south-facing roof areas, and a storage system equivalent in volume to the average garage might be required.

Collector designs are varied, and so are the systems into which they connect. Most systems are water-based, rather than air-based – water has a higher heat capacity and is easier to connect indirectly to the heating system, and water-based systems are also more efficient than air-based ones.

Ventilation and air movement

The effects of ventilation and air flow are important, both around the exterior and through the interior of buildings. There can be either beneficial or negative impacts. Normally, two naturally-occurring forces provide the impetus for air movement:

- the wind effect;
- the stack effect.

In the first, the flow of external wind produces regions of high and low pressure on the building façades, leading to influx and outflow of air through purpose-designed openings (ventilation) and through cracks and gaps (infiltration).

In the second, differences in temperature between the inside and outside of a building produce buoyancy forces which also create pressure differences, which in turn lead to the movement of air within the building, acting like a chimney stack (hence the name). Some buildings are designed to use the sun to deliberately warm columns of air to enhance the stack effect and increase ventilation – such a device is known as a 'solar chimney'.

At present, there is great encouragement to avoid the use of air conditioning by substituting natural ventilation, or mechanical ventilation without cooling, where either is possible. At the same time, excessive heat loss by ventilation in winter periods must be minimized. These two requirements are not easy to satisfy in the same design solution. In the text that follows, we will consider natural ventilation first, before the alternatives. In some cases, such as dense-occupancy office space, it is desirable to maximize ventilation to permit natural cooling in summer months. In other instances, such as dwellings built in cool climates, the primary aim will be to minimize air flow and the associated heat loss in winter.

External wind environment

If the wind effect around the building, and heat loss from it, are to be minimized, then the following design guidelines should be observed:

- Building dimensions should be kept to a minimum to reduce wind pressures.
- The larger building dimension should not face into the predominating wind direction (i.e. the long axis should be parallel to the wind flow).
- Long, parallel rows of relatively smooth-faced buildings should be avoided.
- Tall buildings should have a façade which is staggered and steps back with increasing height away from the wind.
- Sheer vertical faces to tall buildings can generate substantial downdraughts which can obstruct pedestrian access, and may even be dangerous.
- Flat and low-pitched roofs should be avoided, as these tend to increase air pressure differentials.
- Buildings can be grouped in irregular arrays, but within each group, the heights should be similar and spacing between them should be kept to a minimum (no more than a ratio of about 2:1 in building heights).

Wind flow in the immediate external environment has a substantial effect on perception and comfort for those outside in its vicinity, and upon entrance doorways. The shapes and groupings of buildings have a great impact on this, and designers should seek to minimize problems. Wind tunnel testing may be necessary in some cases. In the general areas around structures, the following should be considered.

- Protection for pedestrians can be provided by the use of canopies and podiums which reduce downdraught at ground level.
- Building layout should avoid creating a tunnelling effect between two adjacent buildings.
- Tall slab buildings should not be pierced with walkways or roadways at ground level, since this leads to areas of extreme pressure difference and air flow.
- Shelterbelts – trees, hedges, walls, fences – can provide a degree of protection to both buildings and pedestrians.
- Shelterbelts are most effective when they are

correctly orientated and have a permeability to air flow of about 40 per cent. This level of permeability allows wind to be diffused rather than forced over an obstruction, which would otherwise lead to increased turbulence.

If the aim is to maximize ventilation, then obviously the recommendations above can be reversed. This circumstance is most likely to arise in relation to non-domestic buildings, such as offices, to enhance cooling effects.

Internal air flow and ventilation

Air flow in the interior of buildings may be created by allowing natural ventilation, or by the use of artificial mechanical ventilation or air conditioning. The construction of buildings using more than one of these options is becoming more frequent: such buildings are said to be 'mixed-mode'. Wherever possible, designers must seek to minimize the need for artificial systems, and one way of achieving this is to make greater use of natural ventilation in conjunction with climate-sensitive design techniques for the building fabric.

If air flow is to be encouraged to help provide natural ventilation and cooling, the following are desirable design features:

- The plan form should be shallow, to encourage single-sided ventilation and cross-ventilation.
- Arranging openings on opposite walls to allow cross-ventilation is better than arranging them on one wall or adjacent walls.
- The building depth should not be more than about five times the floor-to-ceiling height if cross-ventilation is to be successful.
- For single-sided ventilation, the building depth should be limited to about 2.5 times the floor-to-ceiling height.
- Minimum opening areas should be about 5 per cent of floor area, to provide sufficient flow.
- External shading devices must be planned carefully, since they can reduce wind-induced pressures and air flow.
- Continuous, secure background ventilation should be available, using trickle vents and other devices.
- Windows should be openable, but able to provide *controlled* air flow (this is particularly difficult in high-rise buildings).
- Atria and vertical towers can be incorporated into designs to allow the stack effect to draw air through the building, though fire and smoke movement regulations may restrict what is possible.
- The need for natural ventilation and cooling can be reduced by using low-energy, controlled lighting and low-energy-use office equipment, thus reducing internal heat gain.

If, on the other hand, air flow inside the building is to be restricted:

- Main entrances should be protected from high wind pressures.
- The provision of entrance lobbies, bracketed by sets of self-closing or automatic doors, helps reduce air influx.
- Revolving doors may be more effective than swing doors in maintaining an effective seal, and unlike pairs of doors which open together, a free flow of air cannot take place.
- Doors and windows to the exterior should be effectively draughtproofed.
- Doors and other openings into stairwells, and other vertical connections through the building, should be sealed to reduce stack effect-induced air flow.

Mechanical ventilation

Mechanical ventilation involves promoting air flow and movement by using fans and air supply/extractor ducts. Such a system may also serve as the heating system in winter, but in its basic form, no cooling system is incorporated, and therefore the lowest air temperature which can be supplied is usually restricted to ambient conditions. Air conditioning involves the cooling of the air using a refrigeration system. More precise control over air temperature and humidity can be achieved in this way, but usually only within a sealed building.

In many temperate climates, the thermal inertia of a building structure, combined with controlled air flow, should be sufficient to avoid excessive overheating except for a few hours each year. If air conditioning is specified, energy use is likely to increase substantially, which will increase carbon dioxide emissions, and a number of the substances used in the refrigeration process also contribute to the greenhouse effect or ozone depletion. Although international treaties are gradually reducing this by restricting the refrigerants which can be supplied as replacement products, there is still a risk.

If natural ventilation is insufficient to create acceptable comfort levels, then mechanical air flow can be employed. As a first stage, it should be used as a supplement to natural flow. Of increasing interest is the application of 'mixed-mode' designs, in which artificial air flow is used at times of the day or year when natural flow is inadequate. Four variations of mixed-mode ventilation have been identified:

- *Contingency* – mechanical ventilation is added or subtracted from the system as necessary.
- *Zoned* – different ventilation systems are provided for different portions of the building, depending upon needs.
- *Concurrent* – natural and mechanical systems operate together.
- *Changeover* – natural and mechanical systems operate as alternatives (but their use often turns out to be concurrent, because of difficulties in zoning or changeover point control).

If mechanical ventilation is to be used to enhance summer comfort levels, the following features should be encouraged:

- Draw external air from the cool side of the building.
- Consider drawing air through cooler pipes or ducts (for instance, located underground) to reduce and stabilize its temperature.
- Ensure supply air is delivered to the required point of use efficiently, to provide the most beneficial cooling effect without inducing uncomfortable draughts.
- Ensure extracted air optimizes heat removal by taking the warmest and most humid air.
- Integrate the use and positioning of the mechanical system with natural air flow.
- In highly-polluted city-centre locations, air filtration down to PM5 (particulate matter over 5 microns diameter) should be considered.
- Consider night-time purging of the building to pre-cool it, using the lowest-temperature ambient air available.

The last of these options offers many potential benefits since the air delivered to the space can achieve a lower temperature than general external ambient conditions. This is particularly the case where cooler night-time air is passed over the building's thermal mass (often the floor slab), enabling it to cool incoming daytime air.

Further 'natural cooling' alternatives to air conditioning are summarized and described in the next section, 'Cooling strategies'.

An increasingly popular option is 'displacement ventilation'. In this case, air at about 1°C below room temperature is mechanically supplied at floor level at very low velocity, usually about 0.2 m per second. This air is warmed by the occupants, computers or light fittings, etc., causing it to rise and be extracted at ceiling level. Air quality and comfort levels are easier to control using this system. However, not all rooms may be suitable for this strategy, so it should be specified only where appropriate.

Air conditioning

Air conditioning systems demand a great deal of energy to run their heating systems, and even larger amounts for their cooling systems. In addition, the rates of air flow required are often substantially higher than with simple mechanical ventilation systems, thus substantial amounts of energy are required to run the fans. This additional energy consumption is not matched by a commensurate increase in comfort. Air conditioning systems are often operated for large proportions of the day, whereas a suitable building design combined with an appropriate environmental control strategy would eliminate the need for them. The excessive use of air conditioning is particularly extravagant in the temperate climate of the United Kingdom.

Of course, there are some circumstances in which air conditioning *must* be employed, and the authors do not advocate its wholesale abandonment. However, its use should be justified by the particular circumstances. Climate-sensitive design can eliminate the need for air conditioning in most instances.

Where air conditioning is deemed necessary, it is likely to be of prime importance in only a fraction of the whole building, therefore the building design should allow for appropriate compartmentalization, with the conditioned area sealed from the remainder of the building.

It is not possible here to offer detailed design guidance, since this is a highly specialized field. What can be said is that basic architectural decisions can greatly reduce the need for full air conditioning. Later sections of this chapter will offer suggestions for alternative design strategies.

Cooling strategies

Several natural or low-energy-use cooling strategies have already been referred to which can serve as alternatives to air conditioning. In terms of general design features, vegetation and planting around the site, together with landscaping techniques, may serve as valuable elements in cooling strategies by modifying the microclimate. Vegetation provides solar shading and also acts as a natural evaporative cooler due to moisture loss through the leaves. Pools, fountains, waterfalls/ cascades, sprays and other water features all add to the evaporative cooling effect.

Chilled ceilings as a method of providing cooling are not necessarily associated with air flow systems. They have two main advantages:

- Thermal stratification effects in a room are reduced.
- A chilled ceiling counterbalances the effect of thermal buoyancy (rising warm air).

The ceiling may be chilled using a refrigerant. The more environmentally benign method is to employ mechanical night-time cooling to pre-cool exposed floor slabs.

A summary of techniques follows, subdivided into reducing heat gain and enhancing heat loss.

Reducing heat gain

Solar protection

- Attention should be paid to planting and landscaping to improve the local microclimate.
- Vertical glazed façades (facing south in the northern hemisphere) are most easily shaded from solar gain.
- External shading is more effective than internal.
- Shading should be compatible with daylight provision and passive solar gain, and where possible, should maintain external views.
- Use heat-absorbing and heat-reflecting glass.
- In traditional Mediterranean buildings, the outer surfaces were painted in light colours to reflect a portion of the heat gain; this practice can still provide benefits today.

Thermal effects

- Insulation materials restrict the heat flow into buildings.
- Simple radiant barriers made from aluminium foil can be installed within building elements adjacent to air gaps to reduce heat flow.
- Radiant barriers towards the outer layers of construction coupled with insulation toward the inner are most appropriate in hot climates.
- Thermal mass absorbs heat gain and introduces a time lag between absorption on outer surfaces and emission to the interior; temperature fluctuations are also dampened.

Internal heat gain

- Reduce heat gain from lights and equipment.
- Consider installing automatic controls for artificial lighting.
- If possible, reduce occupancy levels.

Increasing heat loss

Ventilation and air movement

- Help cool the occupants by increasing air movement during daytime.
- Cool the structure of the building using the cooler air normally available at night.
- Plan the positions of building openings to enhance natural ventilation.
- Investigate the use of wing walls to improve air flow through openings.
- Allow stack effect flow paths to produce ventilation air movement.
- Consider the use of solar chimneys to enhance stack effect air movement.
- Wind towers and wind catchers can be used to provide additional air flow.
- Internal fans – box, oscillating and ceiling types – should be available when air flow is insufficient.

Evaporation techniques

- Air which does not already have a high moisture content can be cooled by allowing water to evaporate into it.
- Direct evaporation occurs when air passes through tree foliage, fountains and across pools.
- If incoming air to a building passes over a dampened surface, or through a spray or damp material across windows, an evaporative cooling effect is produced.
- Direct evaporative cooling is best in dry climates, where average relative humidity at noon in summer does not exceed 40 per cent.
- In the case of indirect evaporation, the air does not come into direct contact with the moisture, but can be allowed to pass through tubes or pipes which have their outer surfaces moistened.

Absorption of heat gain

- Absorption cooling uses natural sources of heat to drive simple absorption refrigeration systems.
- Lithium bromide and ammonia-based refrigerants are most frequently used.
- Heat is removed from the building by air or liquid cooled by the absorption system.

Radiant heat loss

- Radiant heat loss from building surfaces can be improved by adapting the geometry of the building in relation to the sky and other structures.
- Exposed roof surfaces may allow night-time cooling in suitable climates.

Earth cooling strategies

- The temperature of the earth below ground is generally rather lower and more stable than the air above ground.
- The earth is used to absorb heat either by building wholly or partly underground or by passing air through ducts or passages (usually 1–3 m below the surface) prior to supply to the building.
- If below-ground ducts are used, care must be taken to avoid insect or other infestations, and to take account of the water table.

Lighting

Energy-efficient buildings should make as much use of natural light as possible, therefore this section will focus on the use of daylight. Bell and Burt (1995) recommend five aspects of daylighting design:

- the view out of the building;
- the effect of daylight on the appearance of interiors;
- daylight and skylight;
- combining daylight and artificial light;
- the treatment of sunlight.

Lighting is important because of its influence on occupant experience. Until about fifty years ago, the use of windows and the plan form of buildings was very much influenced by the limits of natural light admission. The development of the fluorescent tube lamp, increased environmental noise and poorer air quality all contributed to the expansion of deep-plan, artificially-lit and air-conditioned environments. Only relatively recently have the advantages of natural daylight been recognized once again.

The sections below which deal with windows, glazing and atria design are also closely linked to lighting in buildings, and contain additional guidance on techniques.

Daylight

The principal factors influencing levels of daylight are:

- the orientation of windows;
- the angle of tilt of windows;
- the obstructions to light admission (e.g. nearby buildings);
- the reflectivity of surrounding surfaces.

The following factors should be borne in mind in exploiting daylight:

- Windows provide a view out and time orientation for occupants.
- Occupants more readily accept variable illumination when daylight is the light source.
- Natural light gives true perception of colour rendering and is the norm.
- It would be unusual to expect to supply all lighting requirements using daylight in non-domestic buildings.

Design for daylight

In order to achieve successful daylighting design, the following aspects should be considered:

- The amount of glazing obviously influences the amount of daylight available, but greater window area is not always better – it may simply increase contrast.
- Large windows admit light but also lead to heat gain and heat loss routes, and thus potential thermal discomfort.
- The decision regarding which rooms to allocate to façades should be governed by the activity which will take place within them – these issues should be considered at the building planning stage.
- The amount of sky which can be seen from the interior is a critical factor in determining satisfactory daylighting.
- High window heads permit higher lighting input, as more sky is visible.
- External obstructions/buildings which subtend an angle of less than 25° to the horizontal will not necessarily preclude the use of natural daylight;
- Obstructions greater than 25° to the horizontal may not interfere with daylight if they are narrow.
- If there are many external obstructions, the room depth should be reduced.
- Daylight normally penetrates about 4–6 m from the window into the room.
- Adequate daylight levels can be achieved up to a depth of about 2.5 times the window head height.
- Rooflights give a wider and more even distribution of light, but also permit heat gain, which may cause overheating.
- In general, rooflights provide about three times the benefit of a vertical window of the same size.
- Rooflight spacing should be 1.5 times the ceiling height.
- Where single-sided daylighting is proposed, the following formula gives a limiting depth (L) to the room:

  $(L/W) + (L/H) <= 2(1-R_b)$

  where:
  L = room depth in metres
  W = room width in metres
  H = height of top of window in metres
  R_b = the average reflectance of internal surfaces.

- In non-domestic buildings, the window area should be about 20 per cent of the floor area to provide sufficient light to a depth of about 1.5 times the height of the room.
- Internal reflectances should be kept as high as possible.

Glare

Glare problems arise when very high levels of either daylight or sunlight enter a space, or when there is severe contrast between areas in close proximity. Here are some suggestions on how to avoid this.

- Building and room planning should avoid glare, especially in situations where VDUs are in use.
- Work areas should be screened from direct sunlight, especially low-angle sunlight, and VDUs should be placed at right angles to windows, where possible.
- Contrast glare should be minimized – splayed mullions and light-coloured internal finishes to window walls are to be encouraged.
- Windows should not be positioned at or behind focal points in a room.

Artificial lighting

Energy use for artificial lighting can be minimized by exploiting natural light. When artificial lighting is employed, the following guidelines should be followed.

- The design of artificial lighting systems should not be extravagant, and lighting levels should be designed to be as low as is acceptable whilst still achieving the standard required.
- Task lighting should be used for specific workstations in order to reduce the level of general background lighting.
- Energy-efficient lamps should be specified – usually high-frequency fluorescent or discharge lamps.
- Compact fluorescent lamps should be specified, where appropriate.
- Luminaires giving energy-efficient light distribution should be chosen.
- The control gear (ballasts) required for lamp functioning should be energy-efficient – for example, high-frequency electronic ballasts are up to 20 per cent more efficient than conventional circuits.

The method of controlling lighting systems needs careful consideration to optimize performance. Four types of control are available.

Timed control

In this method, lights are turned on manually, but switched off automatically, according to a specified schedule. Lights near windows should have separate controls. It can produce savings in spaces regularly used by more than two people.

Occupancy-linked control

In this method, sensors (ultrasonic, infra-red, microwave or acoustic) are used to detect the presence of occupants. The lights are switched on for a set period, then switched off if a presence is no longer detected. This type is particularly appropriate for spaces with low occupancy levels.

Daylight-linked control

This method can be used in conjunction with timed and occupancy-linked systems. Photocells are installed to detect when natural light is adequate, at which point, artificial lighting is switched off. Recent developments include the use of dimming controls to avoid an abrupt change as lights are switched off.

Localized switching

This method allows partial illumination of large areas when they are not fully occupied. It gives occupants individual control. Some form of override switching to turn all the lights off when the building is empty is required.

Problems associated with natural and artificial lighting are considered in some detail on pages 107-9.

Windows and glazing

It could be argued that the ideal energy-efficient building would have no windows, thus reducing the twin problems of winter heat loss and summer heat gain. Such a building would be very unpleasant to live or work in, and most studies of occupants' reactions to internal environments place great emphasis on the benefits of external views.

Nevertheless, windows can be thermal weak links when they are used incorrectly. Discomfort arises in summer, not just from the rise in air temperature due to heat gain, but also due to the rise in radiant temperature from the glass surface itself. Radiant effects are further increased if the occupant experiences unshaded sunlight. In winter, cold window surfaces cool the adjacent internal air, which then falls as a result of buoyancy effects, leading to a cold downdraught. This would also be accompanied by low radiant temperatures. Along with the change in temperature, there may well be an asymmetric temperature field, leading to even greater discomfort.

Some of the benefits of windows and glazing have already been described in the section dealing with solar design. This section therefore focuses on alternative, developing technologies which are likely to increase the environmental benefits of glass.

Multiple-layered glazing

The use of single layers of glass in windows is becoming increasingly rare. There are several reasons for this – for example:

- Multiple glazing offers substantially improved insulation.
- Noise insulation is marginally improved by using multiple layers.
- The number of proprietary systems and design options is increasing.
- Building regulations now forbid the use of single-glazing in many situations.

The addition of a second or further sheets of glazing has the effect of trapping a layer of air. Air is a poor conductor of heat, therefore overall heat transfer is reduced – the greater the number of layers, the greater the improvement. However, with every additional layer, the potential for beneficial heat gain and daylighting is reduced because of the increased reflection and absorption of heat and light by each sheet of glass. There is also a significant reduction in light transmittance with each additional layer.

The spacing between each sheet of glass is important – if it is too narrow, the insulation effect will be poor; if it is too wide, convection currents may be encouraged, which increase heat transfer. The optimum glazing spacing for thermal insulation is about 15–20 mm. If heat transfer is to be reduced further, it is most effective to add further layers of glass or to use some form of convection flow-inhibiting material, usually of a cellular structure, within the cavity. Aerogels (see page 35) are one such a material.

Low-emissivity and gas-filled glazing

In multiple-layered glazing components, the air gap created between the panes of glass provides a substantial reduction in heat flow. The remaining heat transfer across the gap is due to radiation and convection. About 60 per cent of the total heat transfer results from the radiative component, so surface coatings which reduce this will improve overall performance.

The emissivity of a surface is a measure of its potential for radiant heat emission. Most glass surfaces have an emissivity of about 0.85; by applying a coating, the emissivity can be reduced to below 0.2, typically of the order of 0.1.

Low-emissivity (low-E) coatings also have the benefit of reducing heat loss by reflecting about 80 per cent of the infra-red heat back into the building. Table 2.1 compares the performance of different types of glazing in different orientations to the sun.

Table 2.1 Effective net U value, taking account of solar heat gain and orientation

Glazing type	U value (W/m2°C) with solar gain		
	South	*East–west*	*North*
Single	2.8–3.7	3.7–4.6	4.6–5.6
Double	0.7–1.4	1.4–2.2	2.2–3.0
Triple	0.0–0.6	0.6–1.1	1.1–2.4
Double with low-E	0.1–0.8	0.8–1.2	1.2–2.4
Triple with low-E	-0.5–0.3	0.3–0.9	0.9–1.6

Metallic oxides are one source of low-E coatings. Some soft coatings must be protected against damage, and can therefore only be used on the internal surfaces of sealed-unit glazing. However, harder coatings are available which have a more robust finish with a wider range of application. It should be noted that if a layer of condensation forms on the low-emissivity coating of the glazing, it loses performance, since water has a high emissivity.

Sealed low-E glazing units may also have a gas filling instead of air. Argon is frequently used; its conductivity is about two-thirds that of air, and a reduction in the overall U value of about 15 per cent is possible.

As an alternative to gas fillings, the sealed units may be partially evacuated, which also has the effect of reducing the heat transfer. The stresses and strains arising from this must be designed for – in particular, the seals must be able to cope with the increased external pressure.

Frame and spacer influence

The type of frame within which a glazing unit is set can have a significant effect on the overall heat transmission. The frame may occupy 10–20 per cent of the overall aperture area, so it restricts the beneficial impact of daylight, and it can provide a weak link in the thermal performance of the glazing. Wood (both hardwood and softwood), plastic, uPVC and metal (aluminium and steel) frames are all in common use. Metal frames are particularly good at transmitting heat out of a building. This escape route should be sealed with 'thermal breaks', which insulate and separate the inner and outer components of the frame. Wooden frames have a naturally lower conductivity, even less than plastic. However, plastic frames have greater versatility, which can be exploited to improve performance. Table 2.2 compares the performance of different frame materials.

Table 2.2 Effect of frame type on heat transfer through glazing units

Frame type	Typical U value (12 mm air gap) (W/m^2°C)	
	Double-glazed windows	*Double-glazed rooflights*
Wood	3.0	3.4
Metal	3.8	4.4
Metal with thermal break	3.3	3.8
uPVC	3.0	3.4

A particular area of risk concerns the spacer used to separate the sheets of glass in sealed units, which can provide a thermal bridge between interior and exterior panes. The spacer can increase heat flow by up to 10 per cent, and can lead to increased risk of condensation around the edges of the frame. The problem can be reduced by the incorporation of insulating spacers; spacers constructed from hollow polycarbonate and rigid silicone are now under development.

Aerogels

Aerogels are used in window technology to improve the insulating performance. Typically, a layer of silica gel is sandwiched between two layers of glazing. The unit is not quite transparent, but somewhat cloudy in appearance. The resulting thermal performance is much improved compared to gas-filled glazing (see Table 2.3). In situations where there is a requirement for background daylighting while controlling heat transmission, aerogels offer some useful options.

Table 2.3 Comparison of typical heat transfer through different glazing options

Glazing type	**U value ($W/m2^{\circ}C$)**
Single	5.6
Double	3.0
Triple	2.4
Double with low-E	2.4
Double with low-E and argon	2.2
Triple with 2 low-E and 2 argon spaces	1.0
Double with aerogel	0.5–1.0

Heat-reflecting and heat-absorbing glazing

These products are usually considered for application in situations where there is a risk of overheating. The cooling load attributable to solar heat gain through windows is a significant fraction of the total load for a building in most temperate climates, and therefore needs careful consideration to avoid the requirement for air conditioning.

Visible light and solar heat gain are both parts of the electromagnetic spectrum of energy emitted by the sun which falls on the earth's surface. The interaction of glazing with light and solar heat has three components:

- reflection;
- absorption;
- transmission.

The proportion of each is affected by the nature of the glazing and the angle of incidence of the radiation. Directly-transmitted solar heat is known as the 'short-wave component'; that which is absorbed by the glass then re-emitted inwards is the 'long-wave component' (note that long-wave radiation is also emitted outwards).

Modifications in the proportions of reflected, absorbed and transmitted radiation can be engineered by changing the glazing system properties. There are several ways of achieving this:

- using 'body-tinted' glass, which increases absorption;
- using reflective coatings, which increase the reflected component, and usually the absorbed component too;
- using combinations of body-tinted and reflective coatings;
- using shading devices.

It must be remembered that a reduction in solar heat gain by these methods inevitably reduces daylight transmission, though some tinting and reflective products are more selective than others. The reflected component can be increased by changing the angle of incidence – the more acute the angle, the greater the reflection. This technique can be employed in certain situations to create interesting aesthetic as well as thermal effects, and usually involves tilting the glazing towards the ground (see Case Study 4.7).

Body-tinted glass is normally available in a range of colours, including grey, green, bronze and blue. The tinting is produced by the addition of small amounts of metal oxides during production, and it is present throughout the thickness of the glass. The effect is to increase the absorption of radiation within the glazing, reducing the directly-transmitted component. The heat absorbed must be dissipated, however, and the glass temperature increases. The warmth of the glass transmits heat inwards as well as outwards – for this reason, the body-tinted layer would normally be installed as the outer of the two panes in a double-glazed unit. Although body-tinted glass has an effect on heat transmission, its aesthetic value has an important role in its specification.

For improved solar heat gain attenuation, reflective coated glass can be used. The coating is applied to the surface of the glass, and in some cases (to reduce deterioration of delicate coatings) it must be installed on the side facing the cavity of a sealed unit, or by applying a second laminating layer.

Reflective coatings are available in a wide range of colours and with a wide range of performance possibilities. It is easier to specify and produce a reflective-coated glass with specific properties for a specific application than with body-tinted varieties. In hot climates, glazing is specified to reduce heat gain, both by direct solar transmission (which can be as low as

10 per cent in some cases) and by conduction. To achieve this second aim, a double-glazed unit with a reflective outer layer is combined with a low-emissivity-coated inner layer to reflect transmitted heat outwards. Avoidance of glare, and provision of some natural light and view are also considerations.

In temperate climates, a balance must be struck between control of summer heat gain and the benefits of winter sun. In addition, higher levels of natural daylight are required. No two situations are identical, and it is important to consider the full range of options before choosing a particular product or glazing system.

Photochromic, thermochromic and electrochromic glass

Each of these terms describes a variety of glazing in which the transmission properties vary in response to conditions. Tremendous possibilities exist for the development of some of these technologies to allow dynamic control of light and heat gain to match building and occupant requirements.

Photochromic glass changes its transmission properties in response to prevailing radiation levels. Small examples have been in everyday use for some years in the form of sunglasses and spectacles. These react automatically, and thus do not permit a high degree of independent control. There are considerable technical problems in scaling up photochromic glass to window size.

Thermochromic glass varies its optical properties in response to temperature variations. Like photochromic glass, it is difficult to control independently, but its response is tuned to variations in environmental conditions, so it can improve thermal and lighting comfort.

Electrochromic glass is the most refined and controllable of the three options. The properties of electrochromic glass can be changed by the application of a small electrical potential. Each pane consists of complex, multi-layered, transparent coatings. The electrical signal alters the transmission of the electrochromic layer, and thus its opacity to daylight and solar heat.

In order to make electrochromic glazing a viable product, improvements continue to be sought in its operation and in the manufacturing process. The contrast between the translucent and transparent states needs to be greater and more clearly defined than is currently the case; the speed of switching in each direction needs to be quicker; its operating temperature needs to be suitable for use in most building situations, and its lifetime needs to be reasonably long.

It seems unlikely that any single product will be able to satisfy the needs of all climates, and a number of different types may be required, perhaps located in different areas of the glazed façade. Energy cost savings alone may not justify the specification of these products at present, even allowing for developments already under way. The requirement for personal control over the immediate environment, especially in cellular offices, may encourage their development.

Shading devices and techniques

Shading devices are normally used to reduce solar heat gain, though they also restrict light transmission. The best design options are those which permit some diffuse light input while shielding from direct sunlight.

Some forms of shading device have already been discussed in the section dealing with passive solar design. In general, external shading is much more effective in controlling solar heat gain than internal alternatives. Table 2.4 indicates the relative merits of employing different shading options.

An interesting example of active/dynamic shading control is the Arab Institute in Paris. Shading devices open and close iris-like patterns of shutters to continually modify the light- and heat-transmission properties of a façade. The aesthetics of the external façade are most impressive, as are the resulting internal effects. However, such dynamic control would not be suited to all situations because, as with the automatic switching of artificial lighting, it can be distracting for the occupants.

Table 2.4 Effect of shading on solar heat gain through double-glazing

Shading	Sun protection	Solar heat gain factor
None	None	0.64
	Light heat - absorbing glass	0.38
	Dense heat-absorbing glass	0.25
	Gold heat-reflecting glass	0.25
Internal		
	Dark-green open-weave blind	0.56
	White venetian blind	0.46
	White cotton curtain	0.40
	Cream linen blind	0.33
Mid-pane		
	White venetian blind	0.28
External		
	Dark-green open-weave blind	0.17
	Canvas roller blind	0.11
	White 45°C louvred sunbreak	0.11

Thermochromic glass also has potential as an external shading medium, since it will only be activated in sunlit summer conditions.

Heated glass coatings

Buildings designed for temperate and cool climates which incorporate large areas of glazing can create discomfort as a result of excessive wintertime heat loss coupled with cold temperatures. Also, there is a high risk of condensation. This problem can be reduced by applying a thin metallic coating to the glazed sheets, or by placing a layer of very thin conducting material across the sheet and passing a small electrical current across the conductive layer, which slightly warms the glass. Comfort is improved, and the risk of condensation is reduced.

This solution is claimed to be 'low-energy', though the energy consumed within the glazing using this heating mode will be rapidly lost to the exterior, and is unlikely to contribute greatly to overall building heating. However, this system is an ideal candidate to be powered by PV cells incorporated in the façade. If this system is to be employed, it must be within very closely defined limits, in which the glazed wall is multiple-glazed and the remainder of the building highly insulated.

Innovative daylighting systems

In recent years a number of novel techniques have begun to be developed with the potential to optimize and exploit the beneficial effect of daylight by allowing its transmission in different ways into the building interior.

Reflective systems

Mirror-based systems can be used to reflect incoming sunlight and daylight deeper into building interiors. Louvres with reflective upper surfaces can be positioned in the building façade at the window opening, and used to shade occupants from unpleasant direct solar radiation while reflecting light onto the ceiling, providing much greater penetration of natural light.

Fixed-mirror systems have drawbacks, however, they cannot track the sun to optimize performance and sometimes distracting patterns of lines of bright light and dark shade are formed. Controlled versions of such louvres need to be motor-driven, as it is unrealistic to expect occupants to adjust them correctly by hand. They also pose operational and maintenance problems.

Despite these difficulties, mirror systems can have benefits. One example is the system installed in the Hong Kong and Shanghai Bank in Hong Kong which tracks the sun.

Light shelves

Light shelves have been in use for some time. Their function resembles the mirror system described above. They are also normally located in the window and provide shade rather like one enlarged louvre element. The light shelves can be located outside the window, inside or on both sides.

Sunlight is reflected from the upper surface of the light shelf into the room interior and particularly onto the ceiling where it provides additional, diffuse light. In overcast conditions, light shelves cannot increase the lighting level; they operate most effectively when direct sunlight is available. Ceilings are usually designed to be higher than normal to optimize their use.

Some degree of control is possible by modifying the angle of the light shelf either internally or externally, or in combination. Low-angle winter sunlight penetration can give rise to glare. Cleaning the light shelves can present difficulties, especially the external type.

Prismatic glazing

Whilst the systems discussed so far rely on the reflection of light, prismatic glazing operates by refracting incoming light. The system consists of a panel of linear prisms (triangular wedges) which refract and spread the incoming light to produce a more diffuse distribution. The view out is substantially restricted, but the system can be used as an alternative to the reflective louvre system, without some of its drawbacks. Glare can also be somewhat reduced. Maintenance is virtually eliminated if the system is installed between the panes of double-glazed units.

Light pipes

Light pipes gather incoming sunlight using a solar tracking system. The light is concentrated using lenses or mirrors and is then transmitted to building interiors by 'pipes'. These can be hollow shafts or ducts with reflective internal finishes or may use fibre-optic cable technology. A special luminaire is required to provide distribution of the light within the building.

The system is heavily reliant on the availability of sunlight, and for critical tasks or areas a back-up artificial light source is required. Although generally inappropriate for the UK because of the lack of consistent sunlight, an interesting example of the technology is to be found in the roof of the concourse of Terminal One at Manchester Airport, UK.

Holographic glazing

Holographic glazing is still under development, which offers potential advantages over prismatic glazing. A diffraction process is also used, but in this case, the light output can be more finely tuned to produce particular internal light patterns.

Innovative daylighting: Design checklist

The following checklist of design issues draws on work by Paul Littlefair (1989) of the Building Research Establishment, Watford, UK:

- A critical factor in achieving a satisfactory outcome is the ceiling design, since it acts as a secondary diffuser.
- In general, ceilings should be light-coloured and uncluttered.
- Floor-to-ceiling heights may need to be greater than the current norm to allow correct functioning.
- Daylight systems cannot produce light themselves – if insufficient diffuse daylight is available with normal windows (say an average daylight factor of less than 2 per cent), any innovative system is unlikely to be beneficial.
- All possible sunlight positions must be considered when designing the system.
- In climates with substantial cloud cover on a significant number of days of the year, systems relying on reflected-beam sunlight are inappropriate.
- Occupants cannot be expected to make constant adjustments to the system. Installed systems should operate well in a fixed mode, should be easy to control automatically, or should require only relatively infrequent adjustment by maintenance staff during each year.
- Good control of artificial lighting is required to produce energy savings by reducing its use.
- The capital costs of using innovative systems must be examined carefully, including the possible need for higher ceilings.
- Systems need to be maintained to operate effectively.

Atria

Atria have become much more common as integral components of commercial building designs over the last twenty years. They appear to provide that most attractive of combinations: an aesthetically pleasing architectural feature which also has the potential to be environment-enhancing and energy-efficient. Their use as a buffer space and as a means of providing natural light and ventilation to the heart of otherwise large, deep and complex buildings has been exploited by many designers, but their function in environmental terms is rarely simple.

Atria were originally defined as open courtyards in the centre of buildings, and they have only evolved into their covered/glazed mode since the nineteenth century. However, it should be noted that the act of enclosing an open courtyard using a roof glazing system reduces the available daylight by about 20 per cent; in extreme cases, with complex roof structures, the loss may be as high as 50 per cent.

It is clear from surveys of recent designs that atria now take many forms. An assessment of their performance is normally carried out by sophisticated and complex environmental simulation programmes. Normally, thermal, lighting and air movement simulations are required, as well as studies to satisfy fire regulations.

Thermal conditions and air flow issues in atria

Air flow is one of the most significant features affecting the atrium environment. A series of questions should be answered in order to define the conditions and requirements:

- What functions does the atrium serve at base level?
- Are exotic plants to be placed in the atrium, and what conditions do they require?
- What conditions (temperature and humidity) can be expected at different levels within the atrium, and what is the influence of the external climate?
- How is access gained from the exterior into the atrium space?

- Is the atrium seen as a buffer space, with some variation in required thermal conditions permissible?
- At each floor level, are openings into the atrium from surrounding offices permitted and advantageous?
- At each floor level, what is the expected air flow direction?
- At each floor level, what other openings allowing air flow are present such as external windows?

Basic decisions regarding whether the offices at each floor level open into the atrium, and the expected flow regime, office to atrium, or vice versa, can have startling effects on the overall performance.

Lighting issues

The shape and form of the atrium also has an important effect on the availability of natural lighting in the spaces adjacent to it. Factors to consider include:

- the degree of transparency of the atrium roof;
- whether the base level of the atrium is designed to be primarily naturally lit;
- whether surrounding offices require lighting from the atrium;
- the shape of the atrium, stepping-back of floors, their width, depth and storey height;
- the surface finish of the surfaces facing into the atrium in terms of colour and reflectance.

Photovoltaic systems

Energy may be converted directly from sunlight into electricity by using a PV cell. Such cells have no moving parts, create no noise in operation, and seem attractive from both aesthetic and scientific perspectives. Power output is constrained by the availability of light falling on the cell, though significant output is still possible with overcast skies.

The development of PV cells is continuing, and there appears to be great potential for the use of modern cells, most recently as building-integrated cladding materials rather than stand-alone arrays, for example, as adaptations of rainscreens, roof tiles and windows.

The advantages of building-integrated systems include:

- clean generation of electricity;
- generation at its point of use within the urban environment;
- no additional land requirements.

As a result, a number of national and international development programmes now exist to help exploit the potential. In several countries, efforts are being made to expand photovoltaic cell use. For instance, in Germany, a government scheme has aimed to provide over 2,000 PV installations in homes. In Japan, a government-supported programme with 50 per cent cost subsidy is expected to greatly increase the use of roof-mounted PV systems, and the aim is to install 70,000 such systems in homes by the year 2005. In the USA, attempts are being made to stimulate the demand for PV systems to help drive down production costs.

Given the abundance of information and advice available, designers should now be able to grasp the opportunities offered by such technologies, which also allow exploration of a range of new aesthetic options for the building envelope.

Types of photovoltaic cell

Modern photovoltaic cells consist of two thin layers of dissimilar semiconducting materials, one known as 'p-type', the other as 'n-type', each with different electrical characteristics. In most common PV cells, both layers are made from silicon, but incorporating different, finely-calculated amounts of impurities. The introduction of impurities is known as 'doping'. When light falls on a PV cell, the extra energy imparted by a stream of photons of light absorbed by the cell permits a flow of electrons to take place and hence provides the electrical potential output. Cells with different characteristics and efficiencies can be created by using different base and doping materials.

Until relatively recently, most PV cells were made from pure monocrystalline silicon, using a slow, difficult and expensive crystal growth process. The cells take the form of a shiny, blue, crystalline cluster. Special surface treatment processes are used to optimize efficiency, and typical maximum efficiencies are about 16 per cent, with a range of 13–17 per cent. However, performance degrades as ambient temperature rises.

Monocrystalline cells are expensive, and this has led to the development of alternatives, which are described below.

Polycrystalline silicon cells

These consist of small grains of monocrystalline silicon. The cells consist of many small, shiny elements. Production processes are simpler and cheaper than with monocrystalline cells, but their polycrystalline nature allows dissipation of the electrical potential at the grain boundaries which reduces efficiency. Recent production techniques have improved peak efficiency to about 10–13 per cent.

Gallium-arsenide cells

Photovoltaic cells using different materials, such as gallium and arsenide, can deliver high efficiencies without the degradation in performance found in silicon cells at high operating temperatures. Efficiencies of over 20 per cent can be achieved, but the cost of such cells has restricted their use to specialist, usually non-building, applications.

Amorphous silicon cells

These cells are normally dull red in colour. They are easier and cheaper to produce than the other alternatives, but their efficiency is lower, and performance degrades from an initial 10 per cent to around 4 per cent.

The next generation of cells is expected to be of the amorphous type, applied in a thin coating to a backing material – so-called 'thin-film' technology. Such cells are not widely used in building applications as yet. However, the entire south-facing facade of the KRK Building in Osaka, Japan, was recently clad in such cells. This included the windows, which still retained 30 per cent light transmission.

Photovoltaic cells are connected together and arranged in arrays for practical purposes. The cells are sandwiched between a layer of toughened glass on the front, and a rear layer which can be made from a variety of materials – for example, glass, Tedlar or aluminium. For building applications, the most popular type is the monocrystalline cell array, which is available from a number of manufacturers.

Energy output

The energy output from a monocrystalline cell varies with insolation level in an almost linear fashion across its operating range. Output is adversely affected by high operating temperatures, with a drop in efficiency from about 12 per cent at 20°C to about 10 per cent at 50°C. Photovoltaic panels would require active cooling in many building situations to maintain maximum output during summer months. Clearly, this is impractical and costly, and at present the drop in efficiency has to be accepted. An alternative is to encourage ventilation of the panels by suitable design of their location and position in order to permit air flow and natural ventilation cooling to the front and, if possible, rear of the array.

The electrical energy produced by PV arrays is in the form of direct current (d.c.), and since most uses of electricity require alternating current (a.c.), an inverter must be employed. Frequently, the supply of electrical energy is not concurrent with demand, perhaps because of occupancy and use patterns. In such situations, two alternatives exist:

- The excess power can be stored in some form of battery.
- It can be fed into the electricity grid (if this is connected to the site).

The first option entails an energy loss in the conversion process, and also requires the provision of a suitable and substantial battery store. The preferred option in most urban situations at present is the grid-connected system, though a sophisticated control system is required to ensure the output matches the grid phase. This also provides a back-up supply when PV generation is insufficient. A major drawback is that the price the utility company pays for the excess PV electricity produced may be only about one third of the price it charges for supplying electricity.

A number of domestic-scale projects have been undertaken, but the potential for increasing use on commercial buildings is very high. It may be easier to justify the cost of PV cladding materials for commercial buildings, and the occupancy pattern means that they are much better suited to the availability of the electrical power produced (see Case Studies 3.3, 3.4, 3.5 and 4.2).

Except in particular situations, PV-generated electricity is not price-competitive with the grid alternative at the moment, but system costs are expected to continue to fall over the next ten years, and to become a much more realistic alternative to conventional supplies.

Cladding systems

As outlined above, one of the most promising areas for development of PV systems is in the form of building-integrated or cladding-integrated options. PV cladding can be considered as a new building material which not only serves as a source of energy, but protects the building from the elements. PV cladding will not be suitable in all situations, such as sensitive historical areas subject to planning restrictions. However, there are many circumstances in which it is appropriate, especially where glass walls are used, such as:

- rainscreen cladding;
- curtain walling;
- roofs.

Rainscreen cladding has been identified as a prime site for exploitation of PV technology. Such a location is particularly useful because rainscreens naturally permit ventilation behind the cladding, which encourages cooling and improves PV performance. The screens incorporating PV arrays are usually tilted away from the vertical to enhance performance, and this also helps shade windows in summer, thus reducing heat gain. Photovoltaic panels can also be used specifically as sunshading devices in their own right.

Curtain walling is another location in which PV cladding might be used. Care must be taken to avoid the visual intrusion of wiring and junction boxes. However, since glass curtain walling is often chosen because it lends a feeling of lightness and provides unobstructed views, PV cells 'deposited' onto the glazing may be used to create their own aesthetic effect. Different colours, densities and degrees of transparency can now be engineered. This developing glass technology is known as 'thin-film technology' in which the surface of glass panels is coated with a PV layer. This is less efficient at converting energy than silicon wafer technology, but is cheaper in terms of production costs. PV curtain walling is a realistic option for retro-fitting to 1960s and 1970s office blocks.

Pitched roofs offer opportunities for mounting PV arrays, and atria roofs, where the view out is often less important than the general light level, also offer potential sites, as do sunshading devices installed around windows.

In general, the best locations for PV cladding arrays are on surfaces which provide some degree of cooling due to air movement or ventilation. The concept of the photovoltaic ventilated façade in which the problem of thermal heat gain is also addressed is a major area of continuing research.

Installation

A number of considerations are relevant when installing PV arrays. The modules are glass-based products, and should be handled with as much care as glass. PV panels produce d.c. electrical output whenever they are exposed to sunlight, so they cannot be switched off. Therefore, during installation they must remain covered until electrical connections have been made. As stated above, a d.c./a.c. inverter is required to convert the output to a more suitable form, and control gear is necessary.

When designing a PV module system, consideration must be given to future shading patterns which might arise from adjacent site development, as well as existing obstructions. It may not be worth installing arrays across the whole of any façade due to the limited period during each day when the cells may be able to generate power. Cell-mimicking cladding is used in such situations for aesthetic consistency.

The direction and tilt of the arrays are important because of the relationship to sun position. Tilt angle can be modified to give the optimum performance at particular times of the year, depending on solar availability and electrical power demand. Optimum year-round performance in the UK climate and latitude is obtained with a tilt angle of about 30–50° to the horizontal, which is about 10–15° less than the site latitude.

Aesthetic considerations

PV cladding materials can now be obtained in different patterns and colours, depending upon the nature of the cells and the backing material to which they are applied. This offers an increasing range of façade options which might be exploited by architects to create particular aesthetic effects. Thin-film photovoltaic systems, where a layer of a coating is applied to glass, seem particularly promising.

Cells in a range of blue shades or red, green or yellow can be manufactured, but these will not be as efficient as standard cells, which are dark-blue. The module background colour can also be chosen to complement or contrast with the cell, or it may be left transparent

Building control systems

Digital control mechanisms, and the availability of system controllers to operate them, have developed rapidly since the 1970s. The incorporation of computers, modern multiple-parameter optimization techniques and intelligent control has enhanced the opportunity to provide very sophisticated environmental control systems in buildings. Often, the environmental data collection and control system is incorporated within an overall building management system (BMS), which also deals with communication networks, security, fire protection, lift operation, occupancy-related scheduling and a number of other functions. Frequently, the system is under the control of a facilities manager.

The portion of the system dealing with energy is known as the building energy management system (BEMS), which may operate autonomously in some circumstances. The control system need not be located on site, and the system could be controlled centrally for multiple building complexes, or for a series of similar buildings in outlying areas.

BMSs/BEMSs are generally designed to control heating, lighting, ventilation and air conditioning systems to manage the internal environment. They can also be used to control more passive features, such as window-opening and shading device position.

The adoption of whole-building systems control has often been accompanied by centralization of decision-making power. As a result, although the building as a whole may optimize its energy and environmental performance to achieve some centrally-defined goal, the ability of the occupants to influence their own environment has been reduced. This has left people with the feeling that if they do not match the typical person profile, they must simply accept the discomfort. Further, there is evidence to suggest that occupants are more willing to accept poor environmental conditions in buildings employing less sophisticated control systems. In other words, if the occupants have the option to influence their own work or living space conditions, they are more tolerant of an imperfect outcome. Further difficulties can arise if the facilities manager lacks either the time or the expertise to understand the sophisticated analysis produced by the BMS, and thus does not fully appreciate problems or the improvements which can be made.

Control of environmental conditions inside buildings is certainly of crucial importance in reducing energy consumption, and it also affects the well-being and efficiency of the occupants. If the occupants do not understand the control system, or the control system does not provide the required environmental outcome, the whole system can be negated. Chapters 6 and 7 examine some of the problems associated with automatic control systems.

Environmental assessment

Environmental assessment of building proposals is becoming much more common. It appears that many groups involved in building design and construction, such as clients, architects, engineers, developers, contractors and prospective tenants, are now ready to accept the value of environmental assessment.

The assessment process is often broken down into particular aspects of design for ease of analysis. Some critics question whether the correct weightings are applied, whether all significant factors have been considered, and whether the outcome gives a fair overall assessment of the building design, particularly given the range of building types. In particular, there seems to be a division between environmental assessment focused on materials/recycling, and that focused on energy-efficiency. This division is particularly strong in refurbishment assessment.

Assessment categories

A recent European Commission document (DGXVII, 1995) suggested potential headings and categories which identify the main areas for consideration when performing an assessment; the checklist below is based upon those headings, and offers a guide to factors which might be considered in the environmental assessment of buildings.

Resource depletion and fuels

- *Raw materials* – Identify the basic component elements for construction.
- *Secondary raw materials* – Identify which raw materials can be or have been recycled.
- *Renewable materials* – Identify those materials which can be replenished by nature.

Global warming

- *Energy use and carbon dioxide production* – Identify how much is associated with the provision of thermal comfort (including the possible additional energy used in air-conditioned buildings).
- *Lighting* – Calculate the carbon dioxide emissions resulting from electrical energy used for artificial lighting.
- *Energy content of materials* – Calculate the amount of energy used in the production of materials.
- *Lifecycle analysis* – Consider the amount of energy required for construction and in use.
- *Transport* – Calculate the pollution associated with traffic travelling to and from work/home/shops.

Ozone depletion

- *Decline in ozone levels in the stratosphere* – Take into account the influence of CFCs, HCFCs, halons, etc.
- *Refrigerant materials* – Consider which types are being used.
- *Refrigerant leak risk* – Estimate the probability of this.

General environmental issues

- *Acid rain.*
- *Deforestation.*
- *Indoor air quality and air pollution* – Consider the volatile organic compounds (VOCs) used in building products.
- *Noise.*

The report also identified four broad categories into which current environmental assessment methods fall:

- those which provide an overall environmental assessment of the performance of the building;
- those relating to the choice of materials;
- those which analyse the energy consumed by the building services;
- those which analyse the embodied energy or lifecycle energy costs of the building.

The Building Research Establishment Environmental Assessment Method

The Building Research Establishment Environmental Assessment Method (BREEAM) has been developed in the UK to provide a user-friendly assessment tool. The procedure is firstly defined by the building type – for example, dwelling, new or refurbished office, industrial unit – and then by type of environmental effect, from global issues to use of resources, local issues and indoor factors, including health effects.

Credits are awarded for meeting certain targets or levels of provision. The credit totals are then compared with threshold values to determine the category of assessment, from 'excellent' down to 'satisfactory'. If any scheme under assessment was considered likely to obtain insufficient credits to merit a 'satisfactory' rating, the assessors would probably advise on improvements, or suggest not proceeding with the assessment.

Energy use is an important component within the assessment, but numerous other factors are considered. The scheme has been successful in attracting a significant proportion of new office buildings as candidates to be assessed. In a competitive market, a good environmental rating can improve letting possibilities.

The list below indicates the sort of building design and operational features which gain credits. The list applies to assessment of new offices; variations on the list apply to other building types. There are three headings under which credits are awarded.

Global issues and use of resources

- Up to 10 credits may be awarded related to the reduction of carbon dioxide emissions caused by energy consumption.
- One credit may be awarded for specifying boilers producing low levels of oxides of nitrogen.
- Up to 7 credits may be awarded for specifying systems and materials to reduce the use and the potential release of ozone-depleting substances like CFCs, HCFCs, halons and insulants.
- Up to 4 credits may be awarded for specification with the aim of reducing the use of non-renewable materials and/or increasing recycling.
- One credit may be awarded for the provision of a storage space for recyclable materials.

Local issues

- One credit may be awarded if the risk of Legionnaire's disease from air conditioning systems is minimized.
- One credit may be awarded for minimizing problems caused by wind flow around large/tall buildings.
- One credit may be awarded for minimizing the risk of noise nuisance to the surrounding neighbourhood.
- One credit may be awarded for buildings which avoid overshadowing of neighbouring buildings.
- One credit may be awarded for specifying toilets which use less water.
- Up to 3 credits may be awarded related to the preservation of ecologically valuable sites.
- One credit may be awarded for providing facilities to encourage cyclists.

Indoor issues

- One credit may be awarded for specifying water systems to minimize the risk of Legionnaire's disease.
- Up to 3 credits may be awarded for improving internal air quality.
- Up to 2 credits may be awarded for minimizing or reducing the health risks associated with certain materials.
- Up to 2 credits may be awarded for providing visual comfort.
- One credit may be awarded for reducing overheating and discomfort.
- One credit may be awarded for providing a comfortable noise environment.

Embodied energy

The energy embodied in a building product is that consumed to derive, construct, manufacture, transport/deliver and install the particular element in a building. It might be argued that the energy costs associated with its demolition and disposal after use in a building should also be included.

The reason for attempting to calculate the value of the embodied energy is to allow full comparison between alternatives. For instance, a building component which helps insulate a building and thus reduce its energy consumption may require large amounts of energy for its manufacture. A simple analysis of the building's energy consumption while in use would not reveal the energy embodied in the product before arrival on site.

For several reasons, it is not easy to specify embodied energy values for building components:

- There are differences in definitions of the components of embodied energy.
- There are significant discrepancies between different sources of data.
- The transportation energy values can be highly significant, though it is not always possible to establish the location of the sources of a product in advance, and thus what transport will be required.
- For commercial reasons, those with information on energy consumption embodied in production are very guarded over its release, which leads to difficulties in making comparisons.

Summary

The intention of this part of the book has been to describe technologies, techniques and products which are important in the design of energy-efficient and environmentally sensitive buildings. Such knowledge enables designers to make well-informed decisions and choose better alternatives when confronted with the task of producing quality buildings.

Not every technique or analysis can be employed in every case, though as was explained in Chapter 1 and will be developed in the Annex, failure to improve on current practice will ultimately lead to severe environmental problems.

Chapters 3 and 4 present case studies of buildings which have placed energy and environmental issues high on their list of design priorities. Readers may judge for themselves the degrees of success achieved.

3: Low-energy housing case studies

Whilst the general trend in housing is towards modest improvements in thermal efficiency, an increasing number of homes are demonstrating that it is possible to make drastic reductions in the use of fossil fuels for space heating. In several cases, the designers claim to have achieved a 'zero-energy' solution embracing all systems. This chapter will review a number of examples.

In Continental Europe, construction techniques vary considerably. Solid masonry walls are common, in contrast to the cavity construction which is mandatory in the UK. This means that the insulation is applied to the external face. Also, floors and internal walls may be of dense construction, providing thermal mass. The following case studies will illustrate how ultra low-energy design can be achieved with different structural techniques.

Case study 3.1: Zero-energy houses, Wädenswil, Switzerland

Architect: Ruedi Fraefel, Gruningen, Switzerland
Project Manager: Dr Ruedi Kriesi, Wädenswil

In 1989, the local town council promoted a project to demonstrate that ultra-low-energy housing was cost-effective. Ten semi-detached houses were built, four aiming at a zero-energy target (see Figure 3.3); they were completed in 1993. Their area of 240 m^2 includes a basement. A main feature of these houses is their substantial thermal mass. In domestic buildings in this climate, the thermal mass principally serves to provide radiant warming over the heating season. Therefore, the full thickness of the structure is utilized, rather than the first 100 mm which is the norm for offices, with their diurnal phases. Much of the thermal mass is provided by the 350 mm in situ concrete intermediate floors.

Figure 3.1
Wädenswil house – solar façade

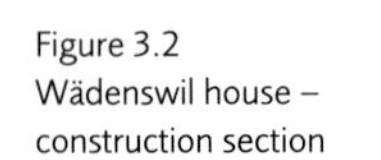

Figure 3.2
Wädenswil house – construction section

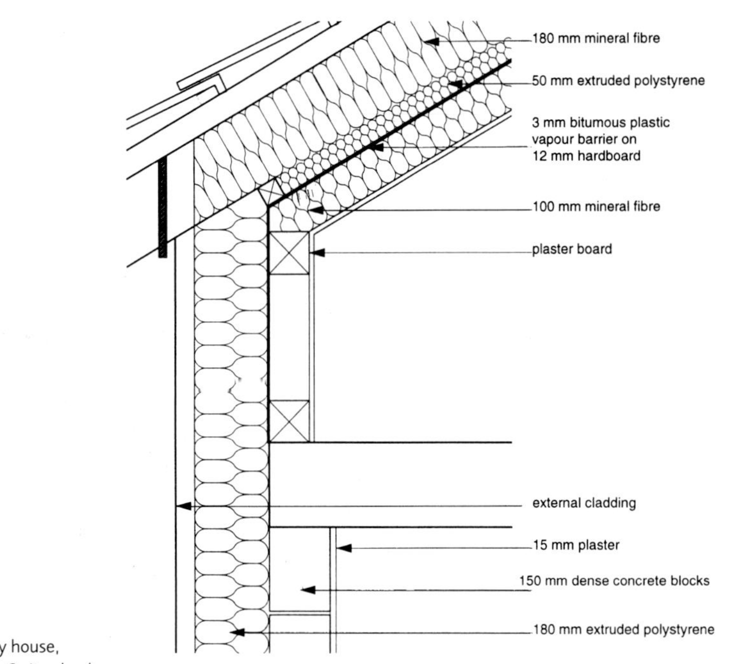

Figure 3.3
Zero-energy house, Wädenswil, Switzerland

Construction

The above-ground walls consist of 150 mm dense concrete blocks, with 180 mm external insulation protected by external cladding. The U value of the walls is 0.15 W/m^2°C. The insulated basement achieves a U value of 0.16 W/m^2°C for walls and 0.19 W/m^2°C for the floor.

The roof is the prime route for heat loss, and a tiled finish covers 330 mm of insulation to produce a U value of 0.13 W/m^2°C.

Windows with a southern exposure are triple-glazed with argon gas filling and two low-E coatings contained in wood frames. Their U value is 1.2 W/m^2°C. North-exposure windows are quadruple glazed with three 25 mm air spaces and two low-E coatings, with an U value of 0.85 W/m^2°C.

Air-tightness is of prime importance at these levels of insulation, and when pressure-tested, these houses achieved 0.4 air changes per hour at 50 pascals (Pa). This necessitated mechanical ventilation to provide adequate air quality, and a ground-coupled system was used, exploiting the constant earth temperature. A construction section is given in Figure 3.2.

Solar radiation is exploited using polycarbonate honeycomb collectors which produce a water temperature of 25°C, even on most cloudy days (see Figure 3.2). Zero-energy-type space heating is provided by pipes laid in the concrete floors, served by the solar collector and supplemented by a heat storage system. The ultra-low-energy type utilizes a back-up liquid petroleum gas-fired unit as its main heat source.

Energy use

Energy use has been measured, and shows an annual figure of 14 kWh/m^2 for the zero-energy house, excluding solar energy. When solar energy is added, the figure becomes 50 kWh/year.

The cost of the ultra-low-energy features was 10–15 per cent above normal Zurich prices. One of the main objectives of the scheme was to demonstrate that ultra-low-energy buildings could be constructed using conventional building techniques.

Case study 3.2: Sixteen terraced passive houses, Kranichstein, Darmstadt, Germany

Designer: Institut Wohnen und Umwelt, Darmstadt

This project is the outcome of several years' research. Its architects, the Institut Wohnen und Umwelt (Institute of Housing and the Environment), used dynamic thermal modelling to calculate the radiant and convective heat flows within the whole house and to and from the thermal mass. The houses were completed in 1995. Detailed monitoring of the houses is under way, using 370 measurement devices. Like Wädenswil (see Case Study 3.1), the aim was to avoid exotic technologies. Figure 3.4 illustrates the house in section.

Construction and design features

Heavyweight construction is again used throughout, with in situ concrete floors (U value 0.16 W/m²°C) and stairs, and 175 mm calcium silicate block walls. The insulation consists of 200 mm of expanded polystyrene under the ground floor slab and 275 mm applied externally to the walls (U value 0.14 W/m²°C). Cement sand-lime render on mesh is bonded to the insulation as the external finish. Triple-glazed windows with double low-E coatings and timber frames give a U value of 0.7 W/m²°C.

The roof is timber, accommodating 400 mm of mineral fibre between masonite I-beams. The roof is covered with turf, giving it a U value of 0.09 W/m²°C.

Air-tightness is exceptional in the last of the houses to be built – 0.2 air changes per hour at 50 Pa. This is compensated for by mechanical ventilation and heat recovery, together with carbon dioxide sensors. Ventilation air is directed mainly to living rooms during the day and bedrooms at night, to minimize the system's electricity consumption.

Space and water heating are delivered from a central source to all the houses, using a condensing boiler. From spring to autumn, domestic hot water is largely provided by wall-mounted solar collectors (solar fraction 66 per cent).

Energy use for the first houses to be built was measured in 1992–3. The predicted level of 31 kWh/m² was remarkably close to the actual consumption of 32 kWh/m².

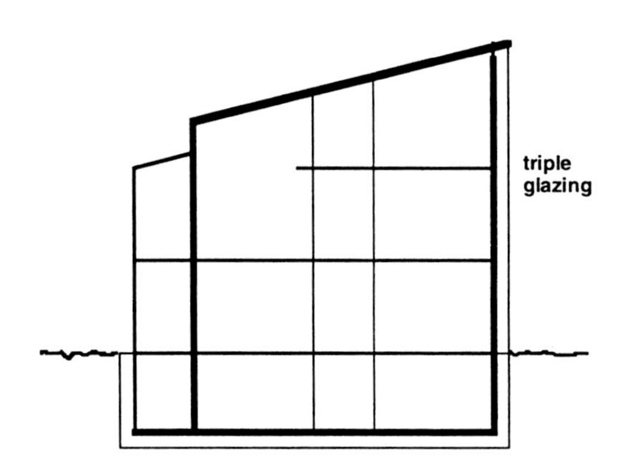

Figure 3.4
Passive house,
Darmstadt – construction section

Case study 3.3: Self-sufficient solar house, Freiburg, Germany

Energy-Design Research and Monitoring: Fraunhofer Institute for Solar Energy Systems, Freiburg. Architect: D. Holken

The objective behind this design, completed in 1992, was to create a house in which the entire demand for space heating, hot water, electricity for equipment and appliances, and energy for cooking is met by solar energy. Full use is made of advanced technology.

A plan form was devised which optimized solar gain, consisting of a segmented/curved southern aspect and a straight, solid wall to the north. Figure 3.5 illustrates the scheme, showing the linking of sections.

Figure 3.5
Self-sufficient solar house,
Freiburg – construction section

Figure 3.6 Self-sufficient solar
house, Freiburg – view of roof

Systems

Highly-efficient solar systems were combined with energy-saving technology to minimize energy demand. A unique aspect of the house is that it stores energy on a seasonal basis in the form of hydrogen and oxygen, which are produced by the electrolysis of water and stored in pressurized tanks outside the building. Hydrogen is reconverted to electricity with a fuel cell, and is used directly for cooking. It also serves as a back-up for heating during cold weather. Short-term electricity back-up storage is provided by lead-acid batteries. Sensors will alert the occupants to any leakage of hydrogen. These have not been activated over 3 years.

The electricity is supplied by roof-mounted PV cells linked to an inverter and a DC net. Figures 3.6 and 3.7 show the building and PV cells. A computerized control and measurement system regulates the operation of the electrolyser and fuel cell. The electrolyser is switched off by active solenoid valves during periods of low insolation, with the PV power then fed directly to the consumers and the batteries.

Domestic hot water is provided by a 14 m^2 curved 'bifacially illuminated' solar collector, with transparent insulation material providing the improved thermal insulation. The hot water it produces is stored in a 1,000 litre insulated tank. The system functions efficiently for nine months of the year, but demand in the winter months, amounting to 360 kWh or 14 per cent of the total, is supplied by the hydrogen storage system via the coaste heat of the fuel cell.

Construction

The elevation to the south, east and west consists of walls made up of 300 mm calcium silicate blocks behind acrylic glass honeycomb transparent insulation material (TIM) (see Figure 3.8). Electrically-operated blinds are installed in the space between the TIM and the external glazing. The U value with the blinds open is 0.5 W/m^2°C, and with them closed it is 0.4 W/m^2°C. The heat gain from the TIM per heating period is between 100 and 200 kWh/m^2. In combination with other conventional means of reducing energy demand, it is claimed that the whole building could be heated solely with TIMs.

Figure 3.7
Self-sufficient
solar house, Freiburg

The storey-height timber windows consist of two independent double-glazed units with low-E coatings, giving a U value of 0.6 W/m^2°C. The north wall is unglazed and is made up of 300 mm calcium silicate blocks externally insulated with 240 mm cellulose fibre protected by timber boarding.

Conclusions

Since 1992, the family of a scientist from Fraunhofer Institute has been living in the house. This provided vital data on the performance of the various systems. There is now evidence that all the passive systems have worked well.

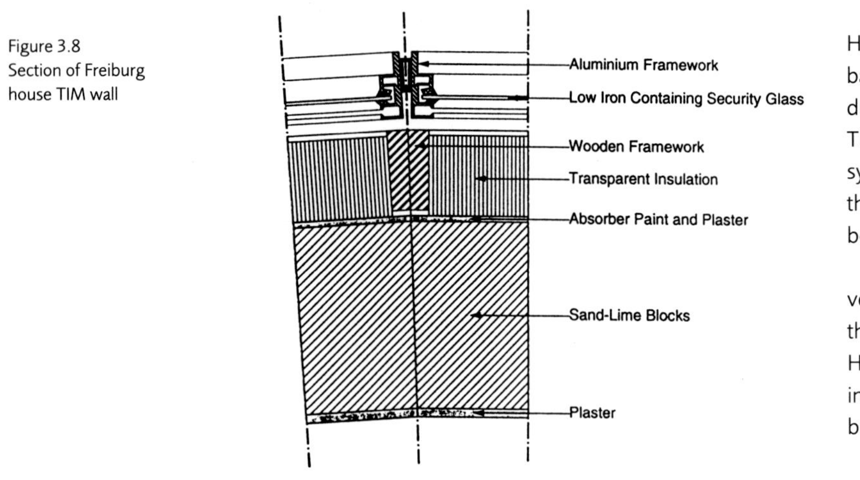

Figure 3.8
Section of Freiburg house TIM wall

However, problems have arisen with the fuel cell and batteries. Extended periods of extreme cold (-10°C), during which there was little heat contribution from the TIM walls, revealed that the original hydrogen back-up system was undersized at 500 W: it should have been three times this capacity. Otherwise, the experiment has been extremely successful within its terms of reference.

On the wider ecological front, its embodied energy is very much on the high side, and the costs were so great that they were only justifiable for an experiment. However, the technology it embodies is likely to be increasingly favoured as pressure mounts to develop buildings and vehicles emitting virtually zero pollution.

Case study 3.4: The Autonomous Urban House, Southwell, UK

Architects: Brenda and Robert Vale

The architects Robert and Brenda Vale are renowned in the UK for ecologically advanced design. Their ultimate achievement is their own house (see Figure 3.9), which they built on a town-centre site close to the medieval Southwell Minster. This alone set them a challenge, since the area was subject to rigorous planning constraints.

Their final design makes an interesting comparison with the Freiburg self-sufficient solar house (see Case Study 3.2), because they also claim the building achieves better than zero demand from the electricity grid. Unlike the Freiburg house, the Vales' house is relatively low-technology: as they put it, others design 'intelligent' buildings, but they design 'thick' buildings. Nevertheless, the building is the product of a high degree of intelligence, since it is designed to be free of almost all services, with the exception of the electricity grid. However, the Vales claim that their PV system makes a net *contribution* to the grid.

Construction

From the outside, the house reflects the character of the town in its form and materials. This disguises the fact that it echoes Continental practice more than the English tradition – for example, it has a basement, which is rare in contemporary English housing. Its first and second floors are concrete, to provide thermal mass. The most striking feature is the level of insulation achieved in all the elements. Substantial block and in situ concrete walls to the basement give a U value of 0.3 W/m²°C,

Figure 3.9
Street view of the Autonomous Urban House

and the floor above, which is within the insulated envelope, has a value of 0.6 W/m²°C. External walls of clay bricks and concrete block inner leaf accommodate 250 mm of mineral fibre insulation, giving a U value of 0.14 W/m²°C. Robust plastic wall ties provide the mechanical link across this exceptionally wide cavity. Local building inspectors had to undergo a radical change of mind set to allow this kind of construction. The Vales were fortunate in having a local authority which sympathized with their aims. In most other localities, their departures from normal practice would have been unacceptable.

Figure 3.10
Autonomous Urban House – garden elevation with PV array

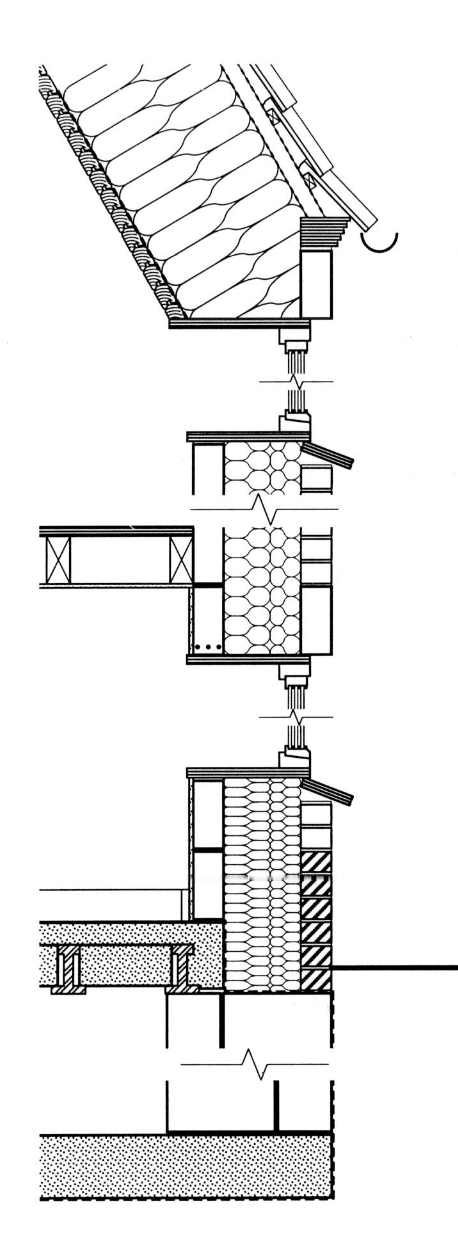

Figure 3.11
Autonomous Urban House – construction section

The roof is another feature whose external appearance hides the fact that its structure is far from conventional. Structural softwood decking beneath a polyethylene vapour barrier and deep I-beam rafters supports 500 mm of cellulose fibre insulation. The clay pantile finish gives the illusion of normality despite a U value of 0.07 W/m²°C. Figure 3.11 depicts the levels of insulation installed in the house construction, while Figure 3.13 shows the interior, which illustrates the exceptional position and thickness of the roof.

The windows are triple-glazed and krypton-filled, with low-E coatings, all within timber frames, giving a U value of 1.15 W/m²°C.

The west garden elevation is dominated by a double-height double-glazed conservatory, with windows opening into the upper rooms (see Figures 3.10 and 3.12).

Figure 3.12
Autonomous Urban House – conservatory

Figure 3.13
Autonomous Urban House – interior view showing roof/ceiling depth

Services

In terms of services, the house uses tried and tested technologies to produce a package which is unique for the UK. A mechanical ventilation with heat recovery system which takes warm air from the conservatory serves the kitchen and the two bathrooms. The bedrooms rely on trickle vents in the windows.

Domestic hot water is provided mainly by a heat pump using exhaust air from the composting toilet. Space heating is provided from the passive solar source of the conservatory. This is supplemented by a 4 kW wood stove with a ducted air supply. It was estimated that this back-up heat supply would need to produce about 3 kWh/m^2 per year to maintain a temperature of 18°C. In practice, the Vales tend to accept a comfort temperature which is below the norm, even for the UK.

Electricity is supplied by a 2.2 kW PV system which is connected to the grid, which serves to even out the peaks and troughs of the intermittent PV supply. The PV cells are mounted on a pergola in the garden. A back-up battery ensures the operation of essential services in the event of power failure. Figure 3.10 shows the PV arrays in position, and Figure 3.14 shows the site plan with the angled pergola which holds the PV system.

Perhaps the most adventurous feature of the house is the water supply which relies on rainwater collected from the roof and fed into 30 m^3 of storage tanks in the basement. From there it is pumped up to a sand filter in the conservatory. The purified water then descends by gravity to a basement holding tank. It is pumped from here to a tank in the loft, to be available as required for general purposes. Drinking and cooking water passes through a further ceramic/carbon filter. There is no mains water connection.

Similarly, the sewage system is also independent of external services. A Clivus Multrum M7 composting toilet serves the whole house. The holding compartment is in the basement and is vented by a 5 W, 12 V d.c. fan, which ensures that the system is user-friendly. Being

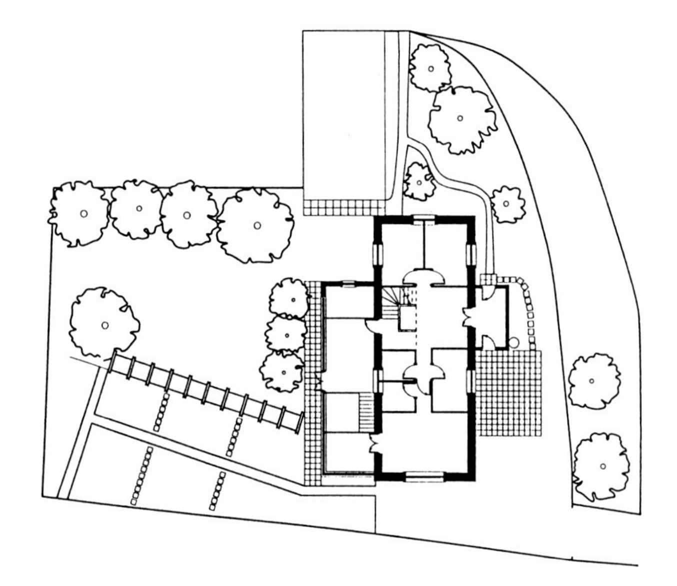

Figure 3.14
Autonomous Urban House – site plan

faced with a seemingly infinite void causes the first-time user some alarm, doubtless for archetypal reasons. The by-product is a permanent supply of garden fertilizer. Grey water enters a soakaway via a grease trap.

Finally, the architects did everything possible to maintain the ecological theme in their choice of materials. For example, the bricks were fired by landfill methane at a works 60 km from Southwell. The hardcore for the slab consists of demolition rubble from the previous building on the site. Internal finishes, furnishings and fittings have been selected with a view to keeping the gaseous emissions from chemicals to a minimum.

Performance

So far, the performance of the house has not been monitored independently. However, the occupants have carefully monitored the energy use, and the output from the PV array between July 1994 and May 1995 was 1,120 kWh, which suggests that there will be a zero net input from the grid. Overall, the calculated energy use is 27 kWh/m^2 per year. Experience so far indicates that the actual consumption will be well below this figure.

As regards space heating, the temperature in the living areas has never fallen below 16°C, with the norm being 18–21°C, and the sleeping areas have varied between 16 and 18°C.

Whilst the domestic water supply has proved more than adequate, the soakaway has been in danger of overflowing several times. The main reason for this seems to be the slow percolation rate of the local soil.

Conclusions

The problems described are relatively minor for a building which has been the realization of the architects' ecological philosophy. From that point of view, it has been a great success – so much so that the local authority has instituted a policy to promote the construction of 100 such houses by the year 2000. Most of these will be funded by housing associations.

From the more general point of view, this building represents the state of the art for energy-efficient housing, now and probably even in the more distant future. What it does prove is that extremely high levels of insulation and air-tightness can be achieved at a cost which is quite acceptable for a quality domestic building. In this case, the overall cost was £550/m^2, including the basement and conservatory, which is well within the range of normality.

Case study 3.5: The Oxford Eco-House, Oxford, UK

Architects: Dr Susan Roaf (concept) and David Woods (detail)

This house was designed and is lived in by Dr Susan Roaf, a lecturer in the School of Architecture at Oxford Brookes University. It is situated in the northern suburbs of Oxford, and is built on an infill site along an ordinary residential road. From the road, one sees its north-facing façade, and it appears to be a new, but otherwise rather traditional, detached house (see Figure 3.15). From the south-facing garden, however, it projects a rather different image, as a result of the solar roof and conservatory.

Figure 3.15
The Oxford Eco-House – view from the road

The accommodation is spread over three floors, in a variation on the conventional pattern. The living, dining and kitchen areas are on the ground floor, the main bedrooms are on the first floor, and further bedrooms are to be found on the third floor, in the attic space behind the south-facing roof. Two bedrooms and a dining room face north; four bedrooms and two living rooms face south. The kitchen and bathroom areas are 'stacked' to reduce pipe runs. The stairs are situated in the centre of the building, and the central portion of the south-facing façade is a double-height conservatory which opens onto living areas and bedrooms.

Care has been taken to use building products with low embodied energy where possible, and to use timber from renewable and sustainable sources. Equal attention to such details was applied to construction materials and finishes and fittings.

The cost of the house is difficult to calculate precisely, as many products were provided directly by manufacturers. The estimated building cost was £200,000 (1995 prices) for a total floor area of about 250 m^2, giving a cost of £800/m^2.

Construction and insulation

Wall construction is of the cavity type, with 150 mm of mineral fibre batt insulation in the cavity. Plastic wall ties have been used to minimize the risks of cold-bridging. The inner leaf is of high-density blockwork, again 150 mm in thickness. Almost all internal partition walls are also 150 mm thick, and the effect is to provide substantial thermal mass to moderate temperature fluctuations. The ground floor incorporates 150 mm of high-density expanded polystyrene insulation, over which is laid 150 mm of concrete, which also adds to the thermal mass effect.

The 'solar-integrated' roof construction includes 200 mm depth of insulation between the rafters (partly using recycled cellulose waste), with an additional 50 mm of polystyrene insulation over the north-facing roof. The south-facing roof external finish consists of a combination of PV panels and flat plate solar collectors, together with two windows for the third-floor rooms. A complex stormproof aluminium roofing bar system is used to hold all items in place. Figure 3.16 shows the rear, south-facing, aspect of the property and the roof.

The windows are triple-glazed, except the buffer spaces (conservatory and front porch) and the rooflights, which are double-glazed.

Figure 3.16
Oxford Eco-House – rear view showing roof

Photovoltaic power

There are 48 monocrystalline silicon cell PV panels (BP Solar type 585) mounted on the roof (see Figure 3.17). They are each rated at 85 W peak output, giving a total output of about 4 kW. Under the north-facing roof is a normal attic void and storage space, which also contains the control gear for the PV electrical system. This allows the d.c. electrical output of the panels to be converted to a.c. for mains use. The mode of use is to export power into the electricity grid at times of surplus, and to import at times of deficit. Unfortunately, the price of electricity bought from the grid is three times the price at which excess power can be sold to it, but since more energy is exported than imported, the net cost over a year is expected to be only about £1.

Over a four-month period during the summer of 1995, an average of 6 kWh of electricity was exported each day, while 2.5 kWh was imported. During the same period, average gas use was 1.7 kWh per day, giving an average energy surplus of 1.8 kWh. A recognized anomaly was that whilst weather conditions were generally good during this period, allowing operation of the PV panels, their efficiency was reduced due to the high ambient temperatures.

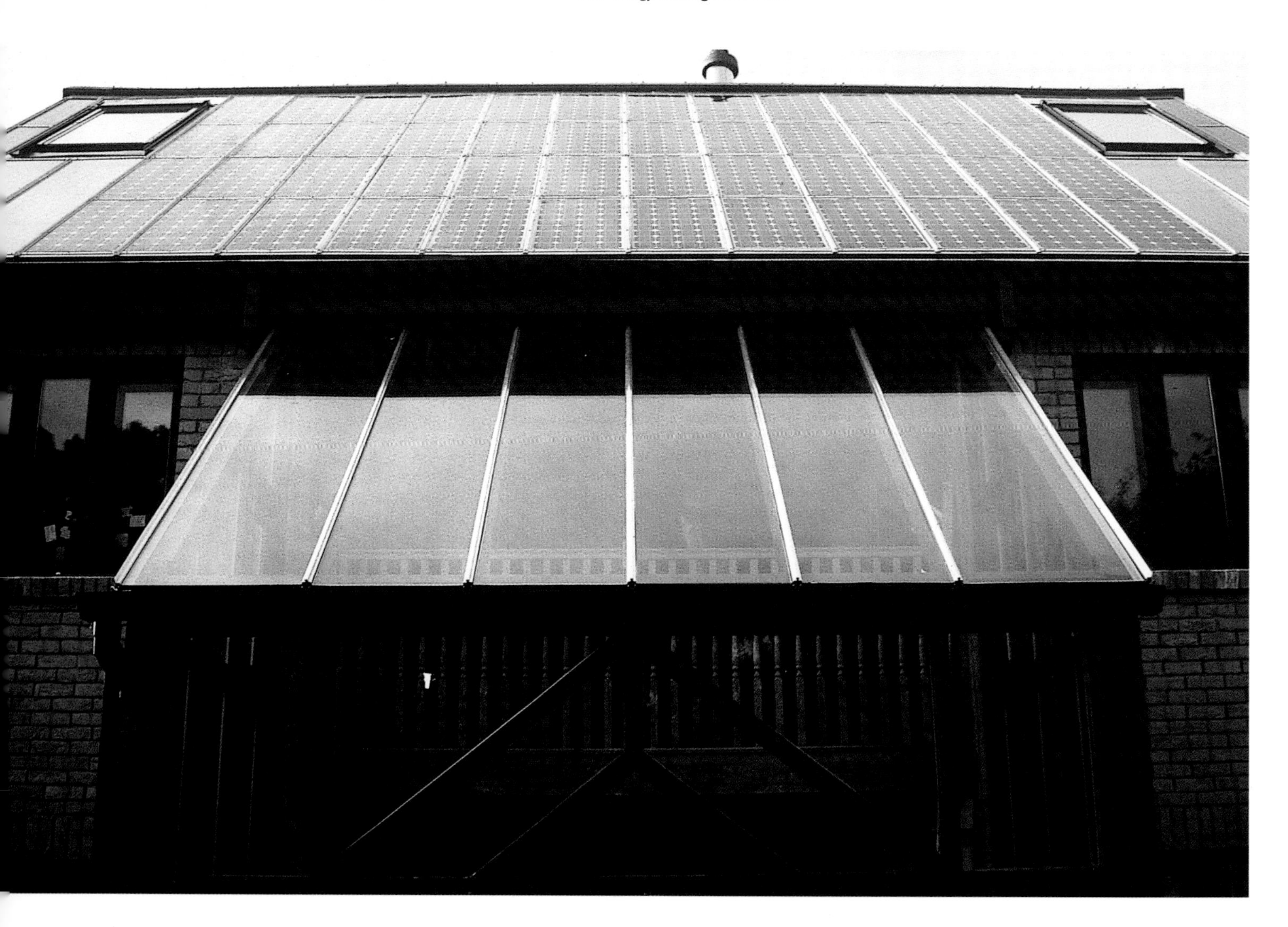

Figure 3.17
Oxford Eco-House – photovoltaic roof

Low-energy electrical equipment is used in the kitchen, and a total of 45 low-energy light sources with a total capacity of 980 W are installed throughout the house. The PV panels can also be used to charge the batteries on Dr Roaf's electrically-powered car.

Heat sources

The house has several sources of heat. A gas condensing coal-effect fire is positioned on one of the internal walls on the ground floor. A hand-built high-efficiency Norwegian wood-burning stove is also situated on the ground floor. Its combustion chamber operates at a high temperature (about 1,050°C) and has a rated efficiency of over 90 per cent. Wood consumption is claimed to be low compared to other wood-burners, and pollutant output is also low. Attempts are being made to avoid thermal stratification in the house by using a fan and ducting.

A gas condensing boiler is also installed, and is fitted with an outside temperature-compensating controller. The boiler serves a total of three vertically-orientated radiators and three towel rails, none of which are on the south side of the house or on the third floor.

The roof-mounted solar hot water panels have an area of 5 m^2, and feed one of the coils in a twin-coil hot water storage tank; the other is fed from the gas boiler. Approximately 70 per cent of the hot water requirements are expected to be provided by the solar panels.

Conclusions

Clearly, a great deal of care and attention to detail is evident in the design and construction of this house – as one might expect of an enthusiast involved in designing their own home. The complete solution encompasses a wide range of energy design ideas. Apart from the south-facing roof, it makes few challenging aesthetic statements, which is perhaps in keeping with the need to attract the general public to the ideas of energy-efficient design.

It has been acknowledged that the cost of PV panels and associated control gear – about £25,000 – would be unlikely to repay the investment within the expected life of the system. This is based on current energy prices and taxes, however, which may alter radically.

The glazing orientation to the south does perhaps provide a single-sided focus to the building, and at least one of the rooms seems to have areas of low daylighting by virtue of its depth. Nevertheless, this is a significant building in that it has facilitated the testing of a number of design ideas. It has been successful in attracting publicity, which is essential if low-energy design is to become the norm.

Summary

Ultra-low-energy homes are still a rarity. As energy costs rise in response to the 'polluter pays' principle, and as the tensions arising from climate change increase, they will surely graduate from being the exception to becoming the norm. What is clear from these examples is that being ultra-energy-efficient does not impose particular aesthetic constraints.

The design characteristics which have featured in all these examples include:

- high levels of fabric insulation;
- high-quality insulating glazing systems;
- use of solar heat gain, in some cases employing novel technologies;
- consideration of reducing all energy consumption, not just space heating and hot water – as evidenced by interest in photovoltaic electricity technologies.

4: Commercial and institutional buildings case studies

Case study 4.1: The Abraham Building, Linacre College, Oxford, UK

Architects: ECD Partnership, London
Quantity surveyors and embodied energy consultants: Davis, Langdon and Everest, London

Linacre College is part of Oxford University; it was founded in the early 1960s and caters for graduate students. As a graduate college, a fairly large proportion of students live off-site but continue to use college facilities. In 1991 it was decided to add extra accommodation to the main site, which is situated not far from Oxford town centre in an environmentally sensitive area on the edge of 'The Parks'. As a result of its location, there were planning and development restrictions. A new entrance/office block was planned (see Figure 4.1), together with an additional 23 study bedrooms. The new accommodation was opened in late 1994, and an official opening ceremony was held in June 1995. In June 1996, the design was named 'Green Building of the Year' in the UK in an award sponsored by *The Independent* newspaper and the Heating and Ventilating Contractors Association.

The main accommodation block (see Figure 4.2) has an area of approximately 1,000 m^2 spread over four storeys, an attic floor and a basement. The building is designed in Queen Anne style to match an existing adjacent block within the college. The four principal storeys accommodate the student bedrooms; the basement contains a gymnasium/fitness suite, and private quiet study booths are provided in the roof attic space (see Figure 4.3). The design brief sought to create an attractive and healthy living environment, an objective which seems generally to have been achieved, with much student praise for the outcome. Some relatively minor difficulties have arisen and these are discussed below.

In order to produce the desired 'green' outcome, a number of design features were specified in the brief. These included:

- low energy consumption;
- high insulation levels;
- use of passive solar design principles;
- ventilation by 'passive-stack' means;
- analysis of embodied energy in the materials of construction and structure;
- materials to be from sustainable sources, either recycled or natural, where possible;
- low associated carbon dioxide emissions;
- use of a recycled grey water system for WCs;
- avoidance of CFCs and HCFCs;
- provision for recycling by residents.

Figure 4.1
Linacre College, Oxford – new entrance and administration block

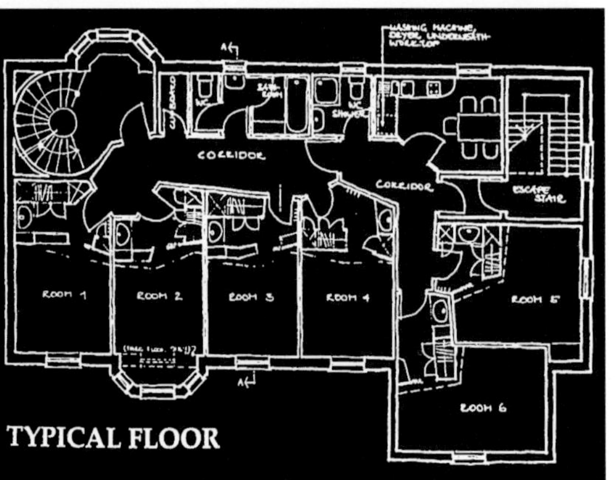

Figure 4.2
Abraham Building, Linacre College – student accommodation block

Figure 4.3
Abraham Building accommodation block – typical floor plan

Energy conservation and environmental features

The building design was constrained by the need to fit, in geography and in style, with other buildings on the site (see Figure 4.4). The complementary aesthetics helped gain planning approval, but also meant that use of such external devices as solar panels would be inappropriate. Also, the payback period for a solar panel scheme was estimated to be equal to its expected operational life and, in any case, did not take account of energy embodied in its construction. Fortunately, many of the building materials chosen for aesthetic reasons could also be made to satisfy embodied energy restrictions. Recycled material was used in the hardcore, roof insulation and roof tiles. Overall, embodied energy is not particularly low, being estimated at 8 gJ/m^2. However, this arises mainly from the construction techniques needed for the basement area.

In-use energy consumption in this building has been reduced by the application of a strategy encompassing high insulation levels and passive solar design which makes use of a shallow building plan with an east–west axis. The majority of the glazing is on the south façade, and most of the study bedrooms face south, with ancillary, circulation and communal areas to the north. Cavity wall construction was preferred, and higher than normal levels of insulation were used, namely 125 mm blown mineral fibre in the walls and 150 mm cellulose fibre in the roof. The construction specification gives a high thermal mass due to exposed concrete floors and masonry walls, and this produces a slow thermal response and dampens temperature fluctuations. Natural or recycled insulation materials were used where possible and, in addition, substantial sound insulation is installed between each study bedroom.

Double-glazed units with low-E glass coatings were employed in the apparently traditionally-styled sash windows, and the window frames are constructed of softwood from sustainable sources. As well as using recycled and recyclable material in the building construction, built-in recycling for the occupants' refuse is provided in each kitchen for glass, paper and metal.

Heating is by condensing gas boiler, and heat distribution is generally by conventional radiators, with thermostatic valves situated in each room. Underfloor heating is installed in some communal areas. A fairly sophisticated control system is employed to optimize performance. Domestic hot water is provided by a packaged direct-fired gas system. A combined heat and power plant was deemed unsuitable for this building because of the scale and cost.

A heat-recovery system is attached to the mechanical ventilation system used in the basement; small cooker hoods and WC mechanical extracts are also used.

Ventilation for kitchens and bathrooms is provided by a non-mechanical passive-stack system which uses 100 mm ducts venting to the ridge of the roof. Trickle ventilators are installed in each bedroom to provide background ventilation. Low-energy lighting is used, together with automatic lighting controls.

A 'two-for-one' tree replacement strategy was included in the planning, as was the sponsorship of a forest area in Tasmania where 19 hectares of eucalyptus have been planted to offset the carbon emissions attributable to the new building's energy use.

The estimated building cost of approximately £800/m^2 in 1993 which took into account the additional energy-efficiency and environmental features, represented a 4 per cent increase over the norm. The building was given an 'excellent' rating using the Building Research Establishment's Environmental Assessment Method.

Figure 4.4
Abraham Building – comparison with the existing accommodation block

Issues for attention

The initial design used the recycling of grey water and rainwater as feeds for WC cisterns. A number of such systems are in use around the world but few in the UK. Unfortunately, problems arose with water quality, causing unpleasant water discoloration and aromas in the cisterns. Initially, this led to the shut-down and draining of the system; however, in the longer term, modifications to the grey water system, possibly incorporating reed-bed technology, were planned to deal with this problem. At current costs it is uncertain whether the system will pay for itself within an acceptable period, though future uncertainties about water supply and demand may make such systems much more attractive.

Low-energy lighting is used in the building and this has given rise to some difficulties. This is in part due to the rather restricted range and the cost of low-energy lighting products and fittings currently available. Spotlights were a particular problem, and some problems with flickering lights were also experienced. Premature failure of the usually long-life, low-energy light sources may also be occurring. The passive infrared detectors used for automatic switching could not be adjusted finely enough for the level of control desired, and may be incompatible with the light sources. There have also been problems with detection because of the limitations on angle and field of view. It has also been reported that the extent of the automatic switching within the student accommodation block means that the occupants are less likely to switch off the remaining manually-controlled lights.

Many of the materials used in construction were specified to be natural or environmentally friendly. The technology to support such products has not yet been as fully developed as for their synthetic alternatives. As a result, care must be taken to select products which avoid premature degradation of organic and water-based sealants. Although the initial intention was to make as much as possible of the building itself recyclable, dismantling has become more difficult because of the reluctance to use lime-based mortars due to compressive strength and setting speed restrictions.

Conclusions

The Abraham Building is clearly enjoyed by its occupants and appreciated by its owners. It appears that the project has been successful in producing an attractive and appropriate building at reasonable cost, and at the same time it has been possible to address global environmental concerns and local environmental issues. The building is a worthy exemplar and benchmark for similar residential accommodation.

Case study 4.2: The Northumberland Building, University of Northumbria, UK

Project team: Northumbria Solar Project, Newcastle, UK; Newcastle Photovoltaics Applications Centre; Ove Arup and Partners, Newcastle; BP Solar, Middlesex, UK; IT Power Ltd Hampshire, UK

The building is situated near to the city centre of Newcastle-upon-Tyne. Serving as a computer building for the University of Northumbria, it was scheduled for recladding and refurbishment. A project team was formed which concluded that here was an opportunity to embark on a more ambitious cladding scheme than would be the norm. The availability of the necessary expertise and the part-financing available through the European Commission enabled the team to give the project a strong environmental emphasis. The building is five storeys high, and the cladding acts as a rain-screen. What distinguishes the project is the incorporation of photovoltaic panels into the south façade.

Figure 4.5
Northumberland Building – before recladding

The cladding panels are designed to visually complement the PV panels and the building as a whole. The project is the largest PV installation in the UK, and the first large-scale demonstration of the technology in a commercial building in the UK. Recladding began in mid-1994, and was completed by early 1995 when the system was officially opened for operation (Figures 4.5 and 4.6 provide 'before and after' views). A number of organizations have been involved in the project, with several objectives:

- to demonstrate the technical feasibility of PV integration in buildings;
- to monitor system performance in a city-centre climate;
- to investigate grid supply–system interactions;
- to accumulate data to aid the development of PV cladding systems.

Photovoltaic system

The building chosen for this project has a south-east-facing façade, oriented 16° from due south. A total of 465 monocrystalline cells are located between the bands of glazing over the five floors of the structure (see Figure 4.7). The cells have a 'laser-grooved buried grid' surface treatment to optimize performance. A notable design feature is the ventilated gap between the building and the rainscreen, which aids cooling of the panels. The array modules are grouped into sets of five within the standard cladding element. Strings of 15 modules are connected in series, each with a nominal output voltage of 270 V. The resulting 31 strings of connected modules feed in parallel into a d.c./a.c. inverter with a power rating of 40 kW.

The cells are positioned at an angle of 25° to the vertical to optimize the capture of winter sunlight; they also provide some summer shading for the windows beneath. The total area of cladding is 390 m^2, 286 m^2 of

Figure 4.6
Northumberland Building – after recladding

Figure 4.7
Northumberland Building – photovoltaic façade

which is the PV panel area. The individual modules have a peak output rating of 85 W, and the total peak output of the system is 39.5 kW. The output from the system supplies power to the building, or if in surplus, to the electricity grid. The peak power so far recorded is 39 kW, and the peak energy output 166 kWh on one day.

The Northumberland Building has a relatively open aspect to the south; however, overshading of some cells occurs (see Figure 4.8), and this can greatly degrade the performance of the whole system. For that reason, some of the cladding modules have been replaced by 'dummy' cells in positions which would otherwise cause problems. In addition, the actual groups of modules connected in each string were carefully chosen to allow control of the output.

The Northumberland Building itself has an electrical load typically in the range 90–120 kW for weekday occupation. The PV system can therefore supply up to about one third of the normal load. Electricity demand reflects the occupancy profile from early morning through to mid-evening. At weekends the load drops to about 40 kW, but this is still in excess of the supply. There is therefore little scope for power export from the building, especially during the winter months.

Project costs

The cost of cladding the building (at 1995 prices) is given in Table 4.1.

Table 4.1 Cost of cladding the Northumberland Building

Rainscreen cladding (excluding PV laminates)	£275/m^2
PV laminates	£423/m^2
Electrical system	£209/m^2
Total cost	**£907/m^2**

Figure 4.8
Northumberland Building, showing potential overshading by an adjacent building

With a total area of 390 m², the total system cost is in excess of £350,000; however, the saving in conventional cladding costs of about £100/ m² must be set against this. The general indications from this project are that PV cladding would amount to 2–5 per cent of total costs for a new building. The cost of the electricity supplied has been estimated to be about 30–40 pence per kWh. This is far in excess of current peak grid supply prices, but one must bear in mind that the scheme was undertaken as a demonstration project, and that future prices of grid-supplied electricity could rise as a result of fiscal measures.

Conclusions

The Northumberland Building makes a bold statement, and has attracted wide-ranging attention in the media. The costs of the PV cladding are considerable, but comparable to polished granite and marble cladding. Therefore, its use for prestigious buildings might be considered by clients or developers in particular circumstances, especially if such buildings could be used to promote environmental issues. Commercial organizations wishing to make a positive statement concerning their commitment to sustainable/renewable energy use might well consider incorporating PV cladding in their buildings as evidence of such a corporate policy.

Advocates of PV systems claim that the technology is proven, that it is environmentally benign and that the systems are commercially available. The Northumberland Building project justifies such claims, though it remains to be seen whether its example is taken up on a larger scale. Certainly, this building illustrates new possibilities for integrating PV cells into buildings, and opens up new avenues for architectural design. However, one must bear in mind that without external support and sponsorship, recladding of this building using PV modules might not have taken place since, at present, if only energy is considered, the cost of the electricity produced substantially exceeds the conventional alternatives.

Case study 4.3: The BRE Energy Efficient Office of the Future, Building Research Establishment, Garston, Watford, UK

Architects: Feilden Clegg Architects, Bath, UK
Services engineers: Max Fordham and Partners, London
Structural Engineers: Buro Happold, UK

The architects, Feilden Clegg, with the backing of the Building Research Establishment as the client, have sought to produce a prototype office scheme which incorporates a number of energy design principles that can be emulated by other architects. It is intended that this 'landmark building' will trigger a new generation of offices that will demonstrate the benefits of integrated passive design to developers and potential clients. The low-energy strategy behind the BRE office makes full use of the building's fabric to reduce or even out heating and cooling loads, and provides a number of alternative ventilation paths in order to accommodate flexibly-arranged office spaces.

Ventilation strategy

The BRE Energy Efficient Office of the Future has been described by the architect as being a 'loose' building. The structure of the building allows a number of natural air flow paths, giving rise to a rather complex ventilation strategy. The 'loose' concept means that passive or natural ventilation will occur whether the office spaces are treated as open-plan or they are made up of cellular units.

The 13.5 m north–south-facing plan is largely cross-ventilated via openable windows and trickle ventilators. Natural wind forces provide air that passes through the office space and back outside through windows on the north façade. Air flow is controlled manually via the lower windows, and by a building management system which controls the top band of windows for high-level ventilation and night-time purging.

Ventilation stacks are also included located on the south façade, the concept being that on a hot, still summer day, when the lack of wind means that cross-ventilation cannot be relied upon, the stacks will heat up, drawing air through the bottom two stories of the building. At present the BMS has been designed so that temperature, not wind flow, will dictate the opening of the stacks. In order to provide back-up to the system and ensure night-time operation, the stack design incorporates mechanical fans. It is also intended that air movement in the stacks will be wind-assisted, as a negative pressure will be created when wind passes across the top of them.

Floor slab

The concrete slab is contoured in a wave form to provide aesthetic appeal and an increased surface area in order to enhance the dampening effect on temperature fluctuations of exposed thermal mass. Voids between the lower curve of the slab and the raised service floor provide paths for ventilation, particularly for night-time air flow, which is drawn from the high-level windows via automatic louvres in order to pre-cool the slab for the following day. Each floor plan is divided into two parts by a circulation zone. On the north side the slab spans the narrower part (4.5 m), and on the south it spans the deeper part (7.5 m). A break in the interlocking slab system occurs above each corridor zone, allowing ventilation paths to by-pass cellular offices which would otherwise prevent cross-ventilation. In other words, air can pass through an office space into the corridor zone and then 'side-step' and vent through the slab over the cellular office.

Cooling

The treatment of the top of the floor slab alternates between raised floors and solid floors. A number of small heating and cooling pipes are embedded within the screed of the solid floors. The scheme hopes to satisfy all of the specified cooling load from a borehole that has been drilled in close proximity to the building site. Water from the proposed 70 m deep hole is fed via a heat exchanger into the underfloor piping system.

Daylight and glare control

In order to reduce the need for artificial lighting, the building is highly glazed on the north and south façades. Detached from the external skin, rotating louvres enable exclusion of direct sunlight and associated glare. Three horizontal bands of louvres shade each of the three storeys independently. The louvres may be adjusted so that they reject summer radiation, admit overcast sky luminance or act as light shelves, depending on the nature of day and the requirements of each storey. Manual control is also permitted.

The louvres are constructed of glass, and this is the first time they have been used in the UK. The glass is coated with a solid white fritted finish with some degree of transparency. The architects wished to avoid the use of solid internal blinds for glare control. In their view, apart from obscuring views out, such blinds make naturally-ventilated spaces feel like parts of air-conditioned buildings.

Figure 4.9
Interior (Courtesy of the Building Research Establishment)

Performance specification

The BRE's office will be the first to adhere to the Energy-efficient Office of the Future (EOF) performance specification. A detailed description of the various elements of the EOF performance specification, which is more specific than the BREEAM rating model, is provided in BRECSU (1995). It is intended that the building's future performance will be monitored and assessed against the targets for energy consumption set by the EOF performance specification.

Summary

The BRE office design uses a range of techniques to improve energy-efficiency. The concept of a 'loose' building is interesting, but it is possible that the BRE design incorporates overcomplicated ideas. In order to prevent the stacks interfering with cross-ventilation, the BMS may need to be programmed so that wind patterns are incorporated into the determination of the size of the openings in the stacks. However, the BRE office does represent advances in passive environmental design concepts, and in the efforts being made to maximize the use of natural resources.

Figure 4.10
Southerly elevation (Courtesy of the Building Research Establishment)

Figure 4.11
Solar baffles (Courtesy of the Building Research Establishment)

Case study 4.4: The Commerzbank Headquarters, Frankfurt, Germany

Architects: Foster and Partners, London

Described as 'the world's first ecological tower', the Commerzbank design attempts to bring the outside environment into the depths of its plan in order to maximize the use of natural resources. Consequently, it is predicted that through the use of simple concepts, together with sophisticated technology, the 60-storey office tower will benefit from unprecedented degrees of daylight and natural ventilation (see Figure 4.12).

Planning strategy

The architects have produced a scheme that is triangular in plan, with curving sides to maximize space efficiency. The triangle shape was chosen to overcome the site constraints imposed by the existing Commerzbank tower. In a conventional tower design, the service and circulation core is positioned in the centre of the plan. The Commerzbank design is innovative in that it has split the core up into three smaller parts and positioned them at the intersection corner points of the 16.5 m wide office floor slabs, allowing the building to possess a full-height central void (see Figure 4.13). This atrium space is connected to the outside by a number of 'sky gardens' that spiral up the tower.

Figure 4.12
Commerzbank Headquarters, Frankfurt – general view

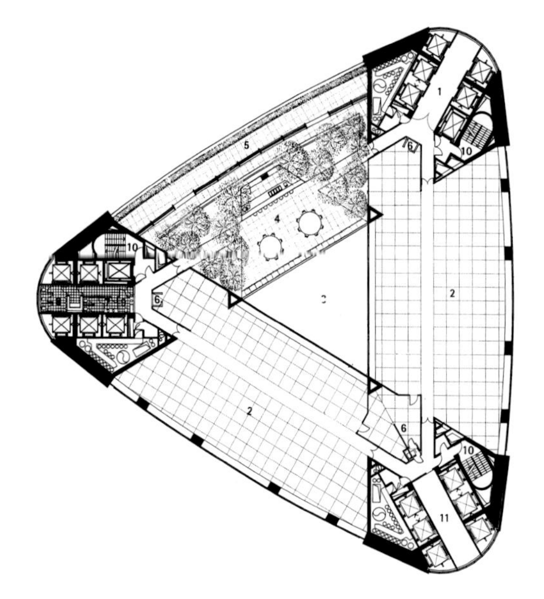

Figure 4.13
Commerzbank Headquarters, Frankfurt – floor plan

Atrium functions

The atrium is subdivided into 12-storey segments, where each segment includes three groups of four-storey-high gardens that rotate around the tower through 120° (see Figure 4.14). The sky gardens, which punctuate the exterior walls of the tower, encourage daylight and fresh air through all three sides of the void over its full height. Wind tunnel analysis predicts that this system will result in the atrium behaving as an outside space, with a high air-change rate.

In order to guarantee air quality, the high-level openings in the glass garden façades are motorized and controlled by the BMS. Depending on the quantity of fresh air that is required in the atrium, the BMS adjusts the size of the openings which are equivalent to a quarter of the 14.5 m garden walls. Therefore, by opening up the interior of the plan to the outside, the design exploits the potential of natural ventilation, as offices on each side of the floor slabs have façades that can be opened to allow the passage of fresh air. In other words, the atrium provides for the same quality of environment for inward-facing offices as those that face outwards, and occupants sitting on the inside of the V-shaped floor plate can gain views across the void through the conservatories to the surrounding city. In addition, there is no solar control through the garden walls, so daylight will flood through into the atrium, vastly reducing the requirements for artificial lighting.

Ventilation

Early research revealed that the largest single achievable energy saving was a reduction in the building's cooling load. Considering a conventional design, and taking into account Frankfurt's climate where temperatures can

reach 35°C in summer, calculations indicated that the building would normally be expected to be in cooling mode for 77 per cent of the year. The most effective way to reduce cooling requirements is to optimize natural ventilation. However, due to the limitations of technology, it was considered an unrealistic target to provide natural ventilation for 100 per cent of the year.

Consequently, the solution is a hybrid or a 'mixed-mode' system, aiming to ventilate naturally for 65 per cent of the year, when air conditioning will be switched

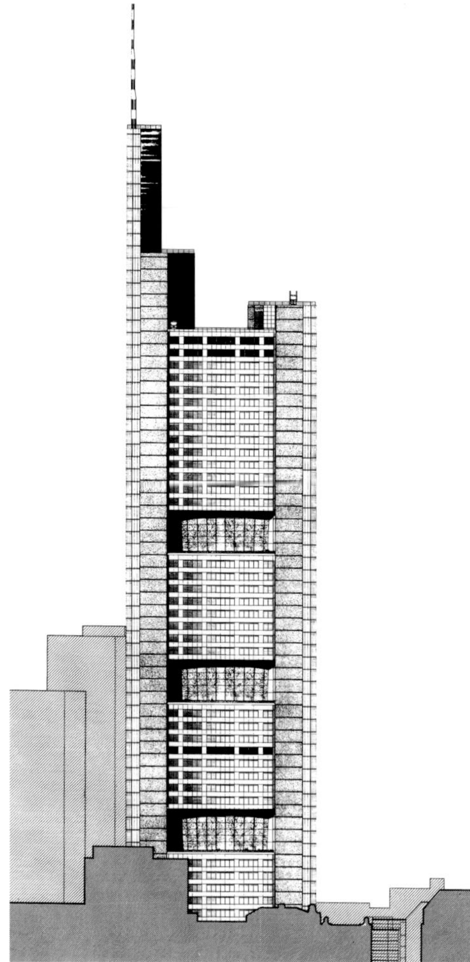

Figure 4.14
Commerzbank Headquarters, Frankfurt – elevation

on for as short a period as possible. The 'cut-off' period, when windows will be closed and the fans will be switched on to operate the chilled ceilings, will occur when the external ambient temperatures exceed 23°C. In addition, the corridor zones, which are located in the central zones of the floor areas, will require permanent air conditioning.

The Klimafassade

The whole thrust of the research relating to the curtain wall design has been to optimize the number of hours the occupants can have their office windows open. Various types of *Klimafassade* ('climate façade') have been analysed in order to ascertain which type will be most suitable for the climatic conditions at the site of the Commerzbank. Through a sophisticated and intelligent design of the *Klimafassade*, the servicing requirements of the tower will be minimized. A number of modelling tests have been conducted with the aim of establishing the most effective combination of openings in the inner and outer skins to promote indirect natural ventilation while also performing as a good thermal barrier. It is possible to introduce a system that performs effectively at a high level in a certain situation, but it may be less effective when conditions differ. Therefore, there needs to be a balanced arrangement that mitigates temperature build-up in summer but traps heat in winter.

After entering locally on each floor through openings in the *Klimafassade* into a 200 mm cavity, air heats up and passes out through the top of the cavity, which acts as a thermal chimney system. The *Klimafassade* construction (see Figure 4.15) consists of 12 mm glazed outer skin that has a special coating to absorb radar signals on the outer leaf. It also incorporates a system of motorized aluminium blinds to give a good shading coefficient, and a low-E double-glazed inner wall with a low U value. Vents are permanently open in the outer skin, whilst the inner section allows the passage of air via a motorized tilt and turn opening arrangement.

BMS control

Occupants can control air flow in the offices by adjusting the opening sizes. However, if it becomes too windy, too hot or too cold, the building management system will close down air flow to individual sections of the building. The different sections are referred to as 'village groups'. By responding to the climatic conditions of each village group separately, the BMS works as a fine-tuned system. Each garden possesses a weather station monitor that feeds appropriate information to the BMS, allowing the system to cater for variations in weather patterns throughout the height of the tower. For example, the BMS might have to respond to heavy wind pressures at the top of the building, but 20 storeys below, the air might be relatively still.

Figure 4.15
Commerzbank Headquarters, Frankfurt – façade

Summary

The design challenges the conventional 'sealed box' solution to high-rise tower buildings. The Commerzbank Headquarters design seeks to enclose high-quality internal spaces whilst greatly reducing energy consumption. The opening up of the plan by introducing an atrium and sky gardens provides the occupants with a stronger connection to the outside, psychologically linking them to nature. The adoption of a mixed-mode ventilation system prevents the scheme from being over-ambitious. For a building of this scale, it is a balanced approach to energy control that will result in the optimization of natural resources throughout its yearly cycle.

The Commerzbank Building represents just one of the possible approaches to passive office tower design. Following its completion, it will be possible to assess the performance of the different energy design features.

Case study 4.5: Learning Resources Centre, Anglia Polytechnic University, Chelmsford, Essex, UK

Architects: ECD, London

It would be reasonable to assume that the design of such a user-intensive building, where up to 700 students occupy 6,000 m^2 of floor space, would rely on air conditioning. At Anglia Polytechnic University, however, the Learning Resources Centre (see Figure 4.20) utilizes passive energy design features such as high levels of thermal mass and air movement through the stack effect to promote natural ventilation. Further credit should be attributed to the project architects, since this environmentally conscious design, which benefits from low capital and running costs, was realized under a 'fast-track' construction process with a restrictive budget.

Natural ventilation

Natural ventilation is promoted by exploiting the stack effect (see Figure 4.16). The top portions of the external windows, controlled by the BMS, pivot open to allow the passage of air, which moves up the atria under buoyancy and is exhausted through vents in the roofs. In order to achieve the optimum stack effect conditions, it was necessary to determine a balance between the proportions of the atria roof openings and the external clerestory vents. The number of windows possessing actuators decreases lower down the building's façade. Consequently, the situation where too much air enters the atria, which would result in the top floor receiving stale air from the floors below, is avoided.

In addition to the higher rates of ventilation required in warm weather, air bricks located beneath the windows act as trickle ventilators to provide a background air movement of 0.25 air changes per hour. External air is ducted through the pre-cast units, drawn into the perimeter casing by the heat from heating elements. Warmed air then enters the space via grilles at the top of the casing. Each duct is damped, providing occupants with the facility to close off incoming air if conditions become too draughty.

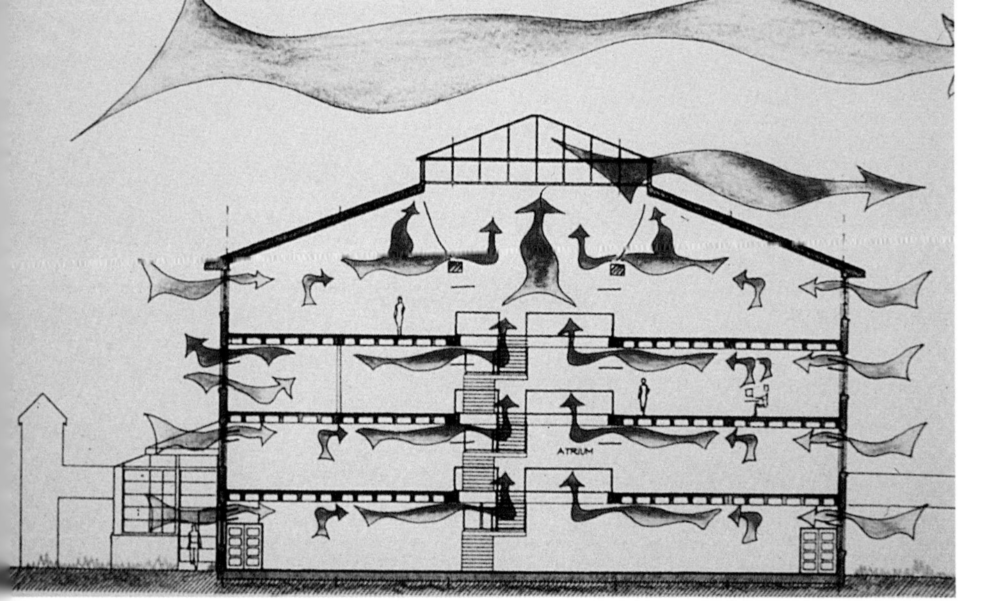

Figure 4.16
Learning Resources Centre – section through atrium

Cooling and thermal mass

It has been estimated that 90 per cent of the building's cooling load will be satisfied by its thermal mass, which absorbs heat in the daytime and dissipates it during the night. After calculating the performance of the thermal mass and the internal temperatures, the BMS determines the period and extent of the high-level window openings for successful night-time cooling.

In order to increase the surface area that is exposed to the air, the concrete floor slab is coffered. Computer modelling has also shown that better contact between the air and the slab is achieved by this design, as air is buffeted between the coffers.

Lighting

In the pursuit of high-quality daylight penetration into the 30 m plan, two atria are included as central design features. The atria are wider at the top, with reflective

Building energy management system

According to the architects, the building has been designed to look after itself; each floor area has a temperature sensor so that the BEMS can regulate window openings. A weather mast is positioned next to the atrium roof, providing the BEMS with wind speed and direction information so that the system can determine which vents to open in the glazed roof. The overall concept is that the Trend software which controls the BEMS will learn, mature and respond to the thermal characteristics of the building. Presently, perhaps as a result of the inherent complexity of the system or lack of operating responsibility, monitoring has shown that lights are left on at odd times.

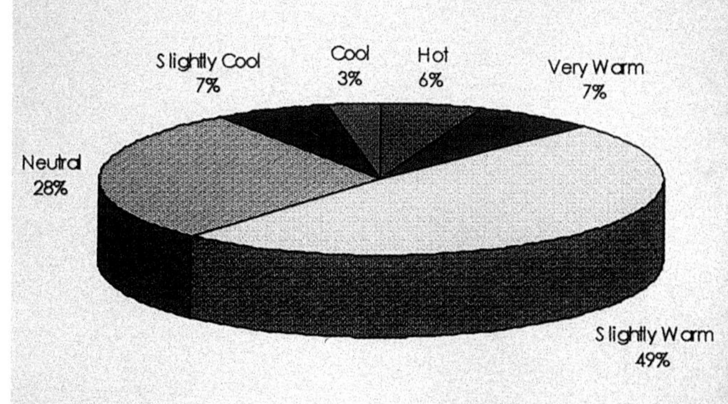

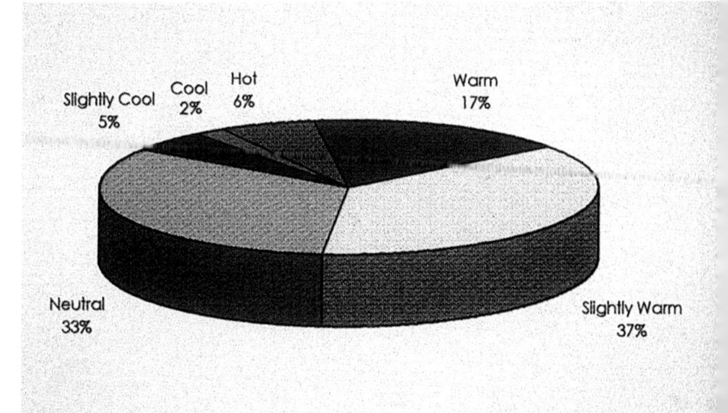

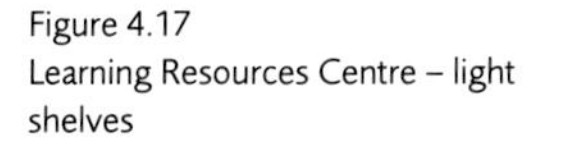

Figure 4.17
Learning Resources Centre – light shelves

Figure 4.18
Learning Resources Centre – occupant thermal comfort in winter

Figure 4.19
Learning Resources Centre – occupant thermal comfort in summer

Figure 4.20
Learning Resources Centre – general external view

surfaces on their perimeters lower down, so that as much natural light as possible is captured and channelled deeply into the building.

Positioned around the top of the atria, glass-fibre fabric sails with a fire-retardant coating further enhance the light and airy internal environment as they reflect daylight downwards. The sails also provide shading as they diffuse the direct sunlight into the top floors.

A double light shelf arrangement is relied upon to mitigate glare (see Figure 4.17). The 700 mm deep light shelves reduce internal gloom by limiting the levels of daylight near the window and reflecting more light into the interior, giving a more even spread of light across the space. Glass has been used for the construction of the shelves so that the view of the sky remains unobstructed for occupants sitting at the window.

Beneath the light shelves, the window has venetian blinds between the two panes of glass in order to minimize peak solar heat gains. High levels of insulation are achieved in the lower 'vision' area, which comprises a double-glazed unit with a low-E coating on the inner pane that reduces the amount of long-wave radiation escaping in the winter.

In order for buildings that rely on passive means to perform effectively, the occupants must understand how the building is intended to operate. Unfortunately, in the case of the Learning Resources Centre, the users' guide produced by the architects and engineers does not seem to have achieved all its aims, and, consequently, the behaviour of occupants may disturb the passive system operation. For example, some users in cellular offices have disconnected the window actuators without realizing that the air flow which had been disturbing them was necessary for the total performance of the building. Nevertheless, studies have shown that the great majority of the users are satisfied with the comfort conditions (see Figures 4.18 and 4.19).

'Design and build' issues

The speed of the design and construction meant that there were only 18 months between the architects' commissioning date and occupation date of the students. Although the Learning Resources Centre scheme demonstrates that passive solar design can be achieved under a 'design and build' regime, the architects had to assume a supervising role at all stages in order to preserve the integrity of their original design. With this kind of procurement, unless the contractors possess a detailed understanding of why certain architectural choices have been made, cheaper alternatives may be proposed, with knock-on consequences for the performance of the building.

In the case of the Learning Resources Centre, frosted glass balustrades were originally specified, with the aim of providing reflected light and preventing cool air from pouring over the edge of each floor and sinking to the bottom of the atria. The construction company apparently convinced the client that open balustrades, formed by metal uprights, would be a more economical option. Fortunately, the architects intervened, and a compromise was reached, taking the form of perforated metal balustrades.

Figure 4.21
Ionica Building, Cambridge

A further compromise was made regarding the nature of the openings on the atria roofs. The architects' preference was for top-hung windows within the slope of the roof. However, the less sophisticated side-pivoting vents that have been chosen, situated only in the vertical faces of the atria, are more exposed to rain, and may perturb the pattern of air flow across the roof.

Modelling predictions and performance monitoring

The Learning Resources Centre has been monitored for more than a year, and has very low energy consumption of as little as one quarter that of a standard air-conditioned building of equivalent area at 114 kWh/m^2 per year, with an estimated reduction in carbon emissions of 82 per cent. Currently, the LRC represents one of eight buildings that is receiving funding through the European Commission's Thermie programme. Co-ordinated by ECD, the monitoring has been carried out by the university's Building Performance Research Unit.

Summary

The high level of sophistication of the Learning Resources Centre design has been achieved through the pursuit of simple technology. An understanding of the passive features of thermal mass and opening windows has produced an intelligent design that has avoided many of the environmental problems others might solve by relying on complex and expensive technology.

Low-embodied-energy materials have been used to create a form that is, on the one hand, robust and 'studentproof', but, on the other, rather predictable and aesthetically neutral. Still, as a result of the efforts that have been made to maximize the use of natural resources, occupants of the centre benefit from an airy and light internal climate.The University is highly satisfied with the building.

Case study 4.6: The Ionica Building, Cambridge, UK

Architects: RH Partnership, Cambridge
Mechanical and electrical consultants: Battle McCarthy, London

This building is located on the St John's Innovation Park on the outskirts of Cambridge, facing the road to Newmarket. It has contrasting façades: heavy masonry on the north, and highly-glazed on the south (see Figures 4.21 and 4.22). With a total area of 4,000 m^2, the Ionica Building encloses a space that has a high degree of flexibility and which is serviced by a mixed-mode ventilation system. The integrated environmental design approach has avoided the creation of a sealed

Figure 4.22
Ionica Building, Cambridge

building, and individual users are able to open windows, allowing a high level of occupant control over the airy and light internal climate.

Ventilation

The overall climatic response strategy involves a three-season approach: the building is ventilated either passively or mechanically, according to the variations between the behaviour of the external climate and the internal comfort demands of each season. In addition, occupants are able to switch between the ventilation systems according to their individual requirements.

During spring and autumn, passive systems control the internal air temperature, windows are opened on all elevations, and the building is naturally ventilated. Air is extracted, aided by the stack effect, via a linear atrium located in the middle of the building, through two types of opening.

The first kind of opening takes the form of six wind towers that punctuate the atrium roof to provide 'close-control' ventilation. Their canopies are designed so that on a windy day, uninterrupted flow occurs across the head of each shaft and a negative pressure is created to pull the air through the opening and close pneumatic doors. The canopies also prevent rain from interfering with the ventilation process. The second opening types also operate under pneumatic control. On a still day, simple incisions in the roof provide a large, open area, enhancing the stack effect ventilation.

The diurnal climate variation is exploited for passive cooling of the building at night. Windows are automatically opened on the south façade, and cooler night air is pulled in and extracted out through the atrium after passing over and pre-cooling the structure of the building. The structure acts as a 'heat sink'; it modifies the internal temperature by being cooled overnight and then absorbing heat generated during the day.

During winter, it is necessary to control the internal air temperature by mechanical means. Air is tempered and modified in order to maintain a temperature of 20–21°C, and is introduced mechanically to the space via a raised floor plenum. The air is then extracted through the central atrium into a heat-exchange unit which is located above the north-facing offices.

Mechanical ventilation is also relied upon during the summer. Air that has been pre-cooled using the heat exchanger and an adapted cooling system is passed through the floor into the space, thus cooling the floor slab and thermal mass of the building. The system employs pre-cast concrete floor planks which are linked by hollow cores so that air can pass through. This is similar in function to the Swedish Termodeck floor design, where cores are cross-connected in each plank to form a continuous, snaking air path, but the Danish version used here is claimed to be more economical, since the Ionica Building permits connections straight through two out of five hollow cores.

Site considerations

The versatility in switching between the ventilation systems caters for the idiosyncrasies of the Ionica Building's site: a refuse dump is situated to the north, and less than a mile away there is a large sewage works. Therefore, as air quality might vary, the ability to close windows and operate mechanical systems becomes a necessity, regardless of the season. The building is further tailored to specific site requirements by the design of the heavy north façade, which has few windows, to provide acoustic protection from the major road at that side. The smaller openings on the north serve compartmentalized office spaces.

Façade design and daylighting

The activities within the Ionica Building rely on a high level of personal computer usage, therefore a major design consideration was to mitigate screen glare while producing an even level of illumination within the space. The building façades have been designed as 'environmental filters': sunlight and daylight louvres are positioned on the south side to control direct radiation and reduce solar heat gain through the façade. Glare control has been introduced by positioning blinds internally. The blinds are controllable in two zones on each window, so that the top third and the bottom two-thirds can be shaded independently.

The 54 m long roof light over the central atrium reduces the effective deep-plan office depth and produces a throw of light onto the internal wall on the north side of the building. In order to prevent glare but at the same time reflect light, the wall is painted off-white. The atrium is covered by a shading louvre system that gives protection from direct sunlight yet allows diffused light into the offices.

Controls

The building management system is organized around a temperature-based decision-making process, and controls the mixed mode ventilation system. Temperature sensors provide information for heating and cooling, and a weather station calculates air speed in order to determine the necessary sizes of openings in the wind towers.

The BMS can be overridden at any point in any season by the user. As the occupant opens or closes windows to control the amount of air flowing through the space, the BMS modifies atrium roof and ventilation tower openings accordingly.

Unfortunately, the subtle relationship between the BMS and the occupant has not been realized to the full. It appears that the problems of occupant/user instructions have not been completely resolved, leaving the staff unable to exploit the climatic versatility of the building fully.

Energy and cost targets

Predictions are that energy usage should be 103 kWh/m^2 per year, and carbon production is estimated at 67 kg/m^2 per year. This would show a large saving in comparison to standard air-conditioned buildings. The system also shows an overall cost saving compared to a traditional air-conditioned building, estimated at £70/m^2 (at 1995 prices) – a figure which should be augmented by reduced lifecycle cost.

Summary

In terms of flexibility for the user in moderating the internal climate, the Ionica Building is unique. Its structure, form and fabric are exploited intelligently to control the environment, with mechanical ventilation employed to temper extreme seasons. The demands of a fast-track construction programme of 18 months from conception to completion meant that decisions and progress had to be made at an unprecedented rate. The process was aided by specialist advice in the areas of daylighting, thermal, air flow, and acoustic design, and by wind tunnel modelling. The design was centred on a sophisticated energy and environmental strategy, and it has led to a robust and architecturally pleasing building.

Case study 4.7: The RAC Regional Control Centre, Bristol, UK

Architects: Nicholas Grimshaw and Partners, UK
Environmental engineers: Ove Arup and Partners, UK

The Royal Automobile Club motoring service's regional office is three storeys high, with an area of 7,000 m^2 (see Figures 4.23 and 4.24). This building is located at the intersection of the M4 and M5 motorways, just outside the city of Bristol. It therefore experiences extremely high levels of noise and pollution, which meant that the use of opening windows to create a naturally-ventilated building was not an option in this case. Although the RAC Regional Control Centre is not a passive building, the scheme displays an environmentally sensitive approach to a difficult site, and incorporates a considerable number of energy design features.

Figure 4.23
RAC Regional Control Centre – façade

Ventilation system

A displacement ventilation system is used in this design. After passing from the plant room and being distributed by three vertical cores, air is supplied through plenums at floor level via a number of supply grilles. The air is extracted through ducts positioned at ceiling level in the ground floor, and through ducts attached to the underside of the roof in the case of the first and second floors, which are volumetrically linked.

The displacement distribution should help satisfy the occupant requirements for fresh air. In addition, it is claimed that the flow of air across the perimeter fenestration reduces the potential for condensation. Heat loss in winter months could increase as a result, though the supply air temperature is kept low, at around 18°C, and in summer months the cooling demand and associated energy consumption is reduced. False ceilings are avoided so that floor slab thermal mass can be exposed in order to dampen temperature fluctuations. This strategy also allows more space for the occupants.

Extract air is taken down to a hygroscopic thermal wheel located in the plant room which is situated in the least desirable space on the ground floor, facing the motorway. After passing through filters which extract dust and pollen, the intake air also enters the thermal wheel, and the wheel transfers heat from the extract air. Thus, heat is recycled but the stale air from the building is exhausted; supply air consists only of fresh air.

Cooling plant

The chillers employed by the RAC building represent the first commercial application of a new range using ammonia refrigerants. Although ammonia chillers cost considerably more than those that rely on environmentally harmful CFCs or HCFCs, their installation may yet prove to be prudent in terms of cost over the system lifetime, since if conventional refrigerant use is further restricted, replacement might be required by law.

Figure 4.24
RAC Regional Control Centre – general view

Floor construction

The lack of suspended ceilings means that each space is bounded by exposed thermal mass, which serves to decrease temperature fluctuations throughout the building's daily cycle. The 18 m span concrete coffered floor slabs act as thermal 'fly-wheels', re-radiating heat during cooler periods. Not only is the coffering of the slabs an efficient structural solution, but the curved forms mean that an increased surface area of the thermal mass is in contact with the surrounding internal environment.

Artificial lighting

The height saved through the use of a displacement ventilation system is used to accommodate suspended high-intensity, low-energy uplighters. The uplighters reflect light onto the floor slabs above, highlighting their interesting coffered form. Light then diffuses into the space below, resulting in an appealing internal ambience through soft background lighting

Daylighting

The RAC employed the architects, Nicholas Grimshaw and Partners, because they supported the concept of an all-glass building. With over 400 people positioned at computer terminals, 'state of the art' measures were necessary to resolve the conflict between the desire to mitigate glare while providing quality daylighting and panoramic views. The result is an elevation which projects outwards at an angle of 16°, the roof overhanging the floor. Horizontal aluminium 'sky limiters' run around the building at each floor level, cutting out views of the bright sky and sky glare. Each perimeter office zone possesses internal motorized roller blinds to further control light admission.

The plan of the building is arranged so that furniture is positioned at right angles to the walls, hence both glare and reflections are avoided at the computer screens. The fenestration comprises double-glazing with a low-E coating and body-tinted glass. In core areas, internal offices are compartmentalized by glazed screens, and receive borrowed light from the central atrium which forms the organizational hub of the building. Figure 4.25 shows a view of the interior.

Figure 4.25
RAC Regional Control Centre – internal view

Building management system

An integrated BMS workstation, controlled by a facilities manager, monitors every zone within the building. Temperature/thermostatic sensors, positioned in 50 m^2 zones, transmit information to the BMS to allow it to control the ventilation and heating of the building. In addition, movement sensors judge whether a space is occupied or not; if there is no movement, the BMS will switch the lights off. As a consequence of the incorporation of such a comprehensive BMS, the users have minimal control over the internal climate.

Site considerations

The building is situated so that two-thirds of its façades face away from the adjacent motorways. The bowl that it sits in, created on a reclaimed site, provides shelter from the road at ground level, and also an opportunity

Figure 4.26
RAC Regional Control Centre – walkway entrance

to position the entrance to the building on the first floor (see Figure 4.26). As a result, the vertical circulation of users is limited to one storey, encouraging the healthy usage of stairs within the atrium or hub, and reducing the need to power the lift.

On the southern corner of the building, a restaurant leads out to a garden, which is sheltered from noise and pollution by the building's structure. Heavy landscaping, particularly in the car park, improves the microclimate. Rows of trees which have been planted are irrigated by an underground reservoir that relies on rain water, which is collected from the roof of the building and stored in a tank.

Conclusions

This building is an appropriate scheme for a case study, since it highlights the compromises which are necessary in certain circumstances, such as the requirement to deal with the high levels of air pollution and noise in the environs of the site.

Case study 4.8: The Elizabeth Fry Building, University of East Anglia, Norwich, UK

Architects: John Miller and Partners, UK
Energy consultant: David Olivier, Energy Advisory Associates, UK

This is an academic institutional building located in the east of the UK (see Figures 4.27–4.30). The most distinctive feature in this design is the reliance on the thermal mass of the structure, supplemented by earth sheltering, to maintain a stable thermal environment. The high specific heat capacity of the earth, together with a number of concrete retaining walls, contributes passively to the slow response of the building by absorbing daytime heat gains and damping short-term temperature fluctuations. In fact, as a result of the heavy floors, roof, walls and internal partitions, the weight of the building is so great that a secondary line of structural columns is required to support a reasonably shallow plan. In order to maximize the surface area of the concrete–air boundary, false ceilings and floors have been minimized. Evidence for the building's unprecedented ability to retain energy was gained when mechanical systems were unintentionally shut down for a weekend during one winter – the internal temperature dropped by less than 1.5°C.

Figure 4.27
Elizabeth Fry Building – south façade

Figure 4.28
Elizabeth Fry Building – north façade

SOUTH

Ventilation

The architects decided that, due to the large number of users (up to 800 people) in an area of 3,250 m^2, mechanical ventilation would be a more effective solution than natural ventilation. Consequently, a system was sought in which the passive features of thermal mass and openable windows would vastly reduce the need for artificial heating and cooling.

In the ventilation of the lecture theatres, air enters the building from a cool external zone on the north side and is drawn into one of the four air-handling plant rooms. It then follows a supply route involving being blown through floor slabs from a duct along the length of the plant rooms. The fresh air mixes with air in an underground plenum amongst a labyrinthine structure beneath the lower ground floor. It then enters the top of the rake of the lecture theatres, drifts down the slope and exits through the light fittings.

The design effectively exploits the potential of night-time purging. Outside air passes through the slabs at

NORTH

Figure 4.29
Elizabeth Fry Building – sunshading

night, cooling the structure for the following day. Possessing cool, radiant surfaces, the fabric should be capable of absorbing heat from up to 1,300 occupants. In order to act as a vast heat sink, the structure has a mean radiant temperature that is lower than average. As a result, comfort conditions can still be satisfied while allowing the internal air temperature to be higher than usual. The necessary degree of cooling is achieved while saving the energy and costs associated with conventional mechanical cooling.

In this building, the structure has been utilized not only as a heat store but also as a distributor for ventilation air through a proprietary floor arrangement. After passing along supply trunking located in the corridor ceiling, air enters the Swedish-designed Termodeck ceiling system. The air zigzags within hollow cores that link within the concrete slabs, and is supplied at a low velocity through grille vents.

Stale air is extracted from the space through ducts that are situated over light fittings; it then returns along the corridors to a heat-recovery unit. The units are also Swedish innovations, being controlled by temperature sensors positioned in the supply and extract ducts. The 'regen air' recuperators have been recorded as operating at high efficiencies – up to 90 per cent heat recovery. In this system, the exhaust air is not re-circulated, and supply air is 100 per cent fresh air.

Construction

The production of a very air-tight construction, especially around the windows to ensure minimum gaps for air leakage, was a prime consideration for the

Figure 4.30 Elizabeth Fry Building – internal view

designers. Post-completion pressure tests have indicated that the high levels of detailing to eliminate draughts and thermal bridging have resulted in a leakage rate of less than 0.1 air change per hour at 50 pascals pressure.

The building has been described as being 'superinsulated', because its structure is wrapped in a large quantity of insulation that is positioned on the cold side of the thermal mass. In addition to the insulation – 200 mm in the cavities, 300 mm in the walls and 150 mm in the floor – the windows are triple-glazed with an argon-filled double unit that has a low-E coating. In the view of the designers, the U values that have been achieved, 0.22 W/m^2°C for the walls and 0.18 W/m^2°C for the roof, approximately twice as efficient as standard building practice in the UK, represent the optimum cost-effective values for energy-efficiency.

Results from first-year monitoring, during which problems with the systems were still being resolved, showed an annual energy consumption of 119 kWh/m^2, which is a substantial reduction compared to the norm for such a building.

Lighting

In order to limit the requirements for the high-frequency fluorescent tube lights, the main circulation space is top-lit. This atrium space encourages maximum daylight penetration, as do the glazed doors in internal corridors, which gain 'borrowed' light. The light sensors that were intended to govern the position of the venetian blinds above the atrium roof have not been fully operational. Fortunately, the blinds are fixed in a advantageous position for daylight penetration (see Figure 4.30).

Conclusions

The building represents a low-energy design solution, achieved within acceptable cost limits, which avoids some of the problems associated with more complex solutions. Extensive passive climate control features, together with high levels of fabric insulation, mean that no BMS is required.

Altogether, the Elizabeth Fry Building represents the antithesis of the advanced-technology approach demonstrated by other case studies.

Case study 4.9: The Low Energy Office Block, British Gas Properties, Leeds City Office Park, Leeds, UK

Architects: Peter Foggo Associates, London
Construction management: Schal International

This building arises out of a regeneration project on a former gasworks site within walking distance of Leeds city centre. It is also strategically placed near the motorway network to the south of the centre. It was conceived as a speculative office development to meet a perceived need for such accommodation in the area. Work carried out with the former Department of Energy's Daylighting and Low-energy Buildings Project helped in the formulation of the design. The commission was won by Peter Foggo Associates, following a limited competition which also involved proposals for the site master-plan for up to 38,000 m^2 of high-quality, flexible office accommodation and 1,100 car parking spaces.

The design which has been produced is the first phase of the development, and comprises a low-rise, low-energy building with a net lettable area of approximately 5,600 m^2 (7,000 m^2 gross area), and parking for 200 cars (see Figures 4.31 and 4.32). The cost of the project was approximately £800/m^2 at 1995 prices.

Figure 4.31
Low Energy Office Block – south elevation

One of the several notable features about its design is that it is a speculative development, at a time when most low-energy building designs in the UK have evolved in response to the requirements of particular clients.

Building form

A strong element of the external design is the landscaping and the interaction of the building with this area. The office block was conceived as a pavilion, with its entrance facing onto the landscaped park. The building is located on the central axis of the site, and is symmetrical in plan (see Figure 4.33). The two 'wings' of office accommodation are slightly splayed and separated by a triangular atrium. The atrium roof is supported by a series of free-standing tree-like structures, and its front elevation consists of a glazed wall with internal sunscreens. The atrium area allows space for displays, social meetings and refreshments.

The building has three principal floor levels, with the car park located partly under the building. The services cores occupy the corners of each office block wing which are adjacent in plan to the apex of the atrium. The staircases, WCs and main risers are all located here. Glazed lifts are also positioned close to this zone, but within the atrium itself.

The low-energy concept has had a strong influence on the external form and elevations. The elevation detail is the same on all façades, and uses a glazed structural cladding system (see Figure 4.34). Windows can be opened manually at two heights. On the external façade, sunshades are set between the high-level-opening vents and the main lower-level-opening lights (see Figure 4.35).

Daylighting and ventilation

The atrium allows the use of natural light and ventilation to be optimized. The stack effect in the atrium is expected to deliver approximately 10 air changes per

Figure 4.32
Low Energy Office Block – internal view

Figure 4.33
Low Energy Office Block – general floor plan

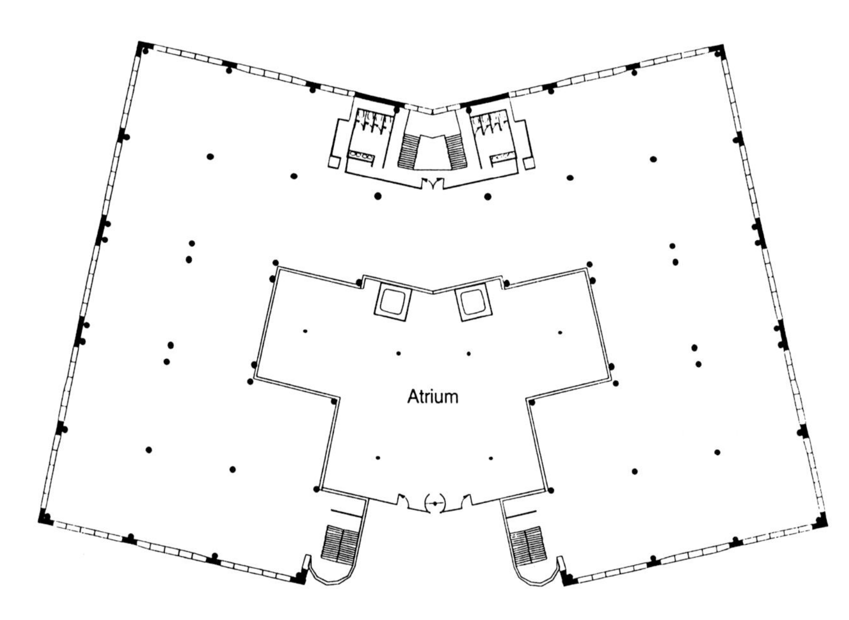

hour during the summer months. The roof consists of glazed and opaque portions which are designed to complement the ventilation strategy by supporting the stack effect which is created, and by the positioning of opening vents. The width of the floor plate has been limited to 15 m in order to provide natural daylight to internal spaces and achieve adequate cross-ventilation.

The windows can be opened in a variety of combinations, with the aim of providing draught-free air movement. The ventilation system operates in a mixed-mode fashion, with fresh air provided via the roof-mounted mechanical ventilation plant. The air enters offices via floor diffusers. In summer, use is made of night-time ventilation, in which air passes through the suspended floor void before entering the building in order to pre-cool the structure in anticipation of the heat gain the following day.

In winter, the heat available at the top of the atrium is used to pre-heat supply air. In summer, the exhaust fan is not used, and the hot air is allowed to vent naturally to the exterior through the openings around the atrium.

The sunscreens perform several functions, providing sunshading during the summer, reflecting daylight into the interior to improve general lighting levels, and serving as external walkways for maintenance purposes. The sunscreen positioning helps to reduce glare and contrast in internal lighting levels that might otherwise result from such a highly-glazed structure.

Upstand structural beams at the perimeter enable windows to reach ceiling height, thereby maximizing the use of daylight.

The artificial lighting installation comprises high-efficiency luminaires and an automatic control system based on perimeter daylight and presence detection.

Thermal design

Exposed concrete has been used for large sections of the structure, and this provides the thermal mass effect which is necessary to enable the operation of the building in a low-energy mode. Particular attention was paid to the soffit detail in order to ensure a balance between structural and thermal requirements, with the ribbed soffit increasing thermal transmittance by approximately 40 per cent compared to a flat slab. The

Figure 4.34
Low Energy Office Block – external façade

Figure 4.35
Low Energy Office Block – external façade showing sunshading

thermal mass helps absorb temperature fluctuations, thus reducing the need for heating and cooling services.

The initial brief for a low-energy design was set by British Gas Properties, although, because of the location, air conditioning was a real possibility. The ventilation strategy, coupled with daylighting and a high level of insulation, creates a purpose-designed control system which avoids the need for it.

The building design does not rely on the opening of windows; comfort conditions can be achieved with the windows closed. There is provision for the addition of air conditioning, should this be required in the future.

Conclusions

The design of a speculative office is often constrained by the need to provide a low-capital cost project which nevertheless offers a high degree of occupant comfort, conventionally perceived as resulting from air conditioning. The design produced here offers a high degree of comfort without air conditioning, yet also provides a low-capital cost and low-running cost building. The environmental controls are quite sophisticated, allowing for the arbitrary opening of windows.

Only when post-completion and post-occupancy monitoring and evaluation of the building has been carried out will a complete analysis of the operational success of the building in achieving the predicted reductions in energy consumption be possible. The designers and developer should be supported in their strategic objectives for this design option, which might act as an exemplar for future design.

Case study 4.10: The new Parliamentary Building, Westminster, London

Architects: Michael Hopkins and Partners, London
Engineers: Ove Arup and Partners, London

Michael Hopkins and Partners were appointed in 1990 to design additional office space for the Houses of Parliament. The design was subject to a number of quite severe constraints. The first was that the building was to be constructed over a new station complex for the Jubilee Line Underground network. Not only did this lead to a delay in starting on site, but it also limited the structure to six points of support inside the perimeter.

The location, opposite the House of Commons, is extremely sensitive in architectural terms, and the new building was required to reflect the massing and site lines of adjacent buildings. The site is also subject to heavy pollution, mainly from traffic. The brief demanded not only a building of highest architectural standards (see Figure 4.36) but also top-quality internal room conditions in terms of air quality, temperature and acoustics.

From the outset, the aim was to produce a building which made minimal demands on fossil fuels. Three principal objectives stemmed from this intention:

- to design the building fabric as the primary internal climate modifier;
- to introduce building engineering systems to assist the building to recycle ambient energy;
- to design the building services in such a way as to greatly reduce their parasitic energy use.

One prerequisite was that there should be a sealed façade on account of the noise and pollution. The façade design was to make full use of the passive abilities of the building's materials and form to maintain indoor climate. Building services were chosen which would enhance these abilities and introduce energy harvesting and recovery. The extended design time allowed the architects and engineers to undertake detailed research not normally possible in these days of severely time-constrained contracts, as, for example, in the case of the Anglia Polytechnic University Learning Resources Centre (Case Study 4.5). This research began with an analysis of the major ambient energy flows entering the building, how they could circulate through it, and how they would leave the building. The architecture and engineering services were developed to complement these flows.

A further constraint was that the building fabric should have a minimum life of 120 years. Accordingly, the durability of materials, their maintenance costs, the replacement strategy and effective use of space all became decisive factors. With a design life exceeding 120 years, the embodied energy becomes insignificant compared with the 'in use' energy over the lifetime of the building. The resulting design predicts an energy consumption of 90 kWh/m^2 per year, based on a 50-hour week. This included ventilation, heating, cooling, lighting, office equipment and miscellaneous power use. This is lower than most of the contemporary generation of naturally-ventilated buildings, and is about 25 per

Figure 4.36
New Parliamentary Building, Westminster – general view

cent of the Best Practice target. Because of the holistic design of the services and the fabric, the building is able to maintain internal conditions at 22 ± 2°C without the assistance of mechanical refrigeration.

The objective of maximizing natural resources involves the following tactics:

- solar heat intercepted by the glazing system;
- night-time outside air cooling in association with building fabric thermal storage;
- ground water as a direct cooling source;
- solar heat directly into rooms;
- internal heat gains;
- daylight;
- daytime 100 per cent fresh air cooling;
- buoyancy-assisted room and duct air flows;
- internal moisture gains.

Figure 4.37
New Parliamentary Building – cutaway detail of the façade

Ventilation system

The mechanical ventilation system serves a network of linked floor plenums throughout the building via ductwork in the façade to provide 100 per cent outside air ventilation to each room.

High-efficiency heat recovery is the key benefit of mechanical ventilation over conventional natural ventilation. Not only does it allow generous year-round ventilation using outside air as opposed to the higher supply temperatures needed with displacement ventilation, but it also permits the recovery of solar heat from the window system, the occupants, the electrical equipment and the room radiators. The latter consist of rotary heat exchangers operating at efficiencies of more that 85 per cent. They are of the hygroscopic type, able to recover winter moisture from the exhaust air and thus reduce the load on the supply air humidifiers. Ventilation ducts are positioned alongside windows and are expressed architecturally on the façade (see Figure 4.37). The extract takes room air through the light shelf and lower window venetian blind cavity via the clerestory glazing plenum. The supply air is fed into a raised floor plenum.

Displacement room air movement is used because, unlike conventional systems, it is compatible with direct cooling by means of ground water at 14°C. This form of room air distribution, together with the vertical façade ducts, allows buoyancy to work in conjunction with minimum fan power. Also, passing supply air through the floor void improves the cooling capacity of both ceiling and floor. The selection of very low-pressure-loss air handling and duct system components means the ventilation energy use target is 1 W per litre of air supplied. The same full fresh air system is able to serve all different room types, thus allowing for future changes of room function without having to modify the services. Not only is this consistent with the long life requirement of the building, it also greatly reduces the embodied energy content of the engineering services throughout its life.

Exhaust air is carried by ducts expressed externally in the steeply-pitched roof, and expelled through a series of chimneys designed to enhance the stack effect and incorporating a thermal wheel (see Figures 4.38 and 4.39).

Heating

For most of the year, a significant proportion of the building will be able to exploit internal heat gains from occupants, machines, lights and beneficial solar gain. These gains will more than satisfy the fabric heat loss. This means that heating the supply of outside air becomes the dominant heating demand. Consequently, the ventilation system design centres on the capacity to recover heat from all the internal heat sources and the window solar collectors, thereby allowing heat recovery to cater for most of the ventilation heating.

The heating system is a variable-water-volume system with thermostatic valves on the room radiant panels, allowing the system to respond to beneficial internal and solar gains, and occupant temperature trim controls. The water flow temperature is 70°C, with a 50°C return temperature to maximize flue gas condensation efficiency in the condensing gas boilers.

Cooling

The requirement for a temperature range of 22 ± 2°C for occupied rooms using passive cooling needed a detailed understanding of the heat flowing into and out of each room. For most of the year, there is an excess of heat to be managed. This excess heat is stored, first to deal with the night-time heat loss and to avoid the need for boost heating first thing in the morning prior to occupancy, and then to allow night ventilation to remove any surplus from the building. High-thermal capacity room surfaces are used as the heat storage medium, with their ability to function with small temperature changes and to take full advantage of both radiated and convective heat transfer.

The room thermal capacity handles the internal heat gains, but for ventilation when the outside air is above 19°C, ground water at about 14°C is drawn from two on-site boreholes to cool the outside air down to room temperature. A displacement ventilation system is used to allow this cooling to be achieved without mechanical refrigeration.

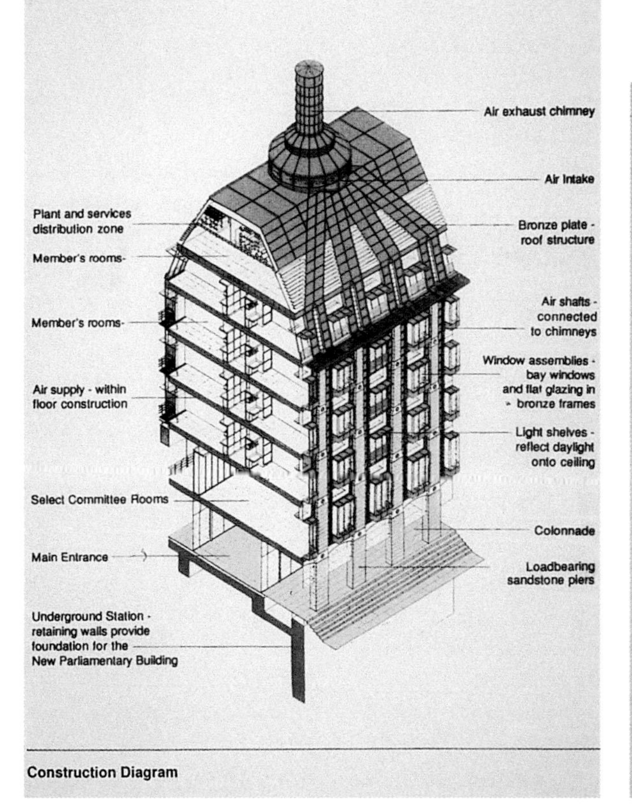

Figure 4.38
New Parliamentary Building – axonometric section

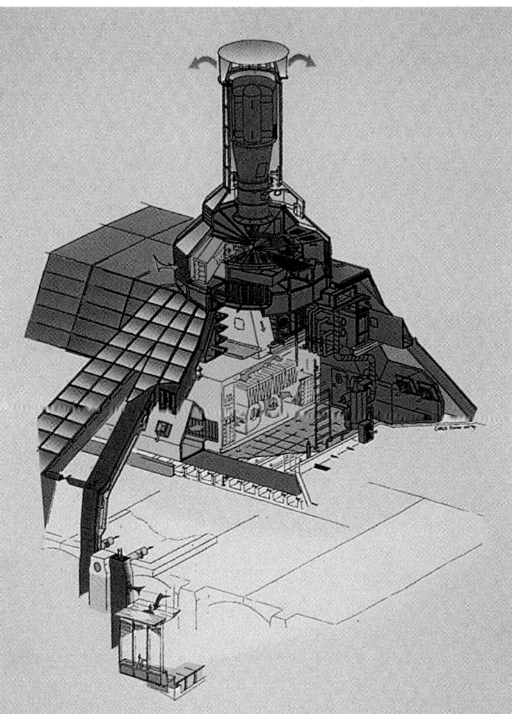

Figure 4.39
Ventilation chimney incorporating a thermal wheel

The façade

The cladding system offers an integrated solution to the provision of external views, room daylight control, passive and active solar energy collection, excess solar heat protection, minimizing room heat loss, ventilation supply and extract, and heat recovery.

Windows consist of triple-glazing, with mid-pane retractable ventilated blinds optimized for colour and emissivity to absorb solar heat. The outer double-glazed unit has a low-E coating and is argon-filled. The cavity is ventilated with a proportion of room extract air, and at the same time acts as a solar collector. This arrangement results in less than 25 W/m^2 summer solar heat gain across the floor area of a 4.5 m deep perimeter room.

The glazing incorporates a light shelf to maintain daylighting when solar shading is in use, thus avoiding the all too common 'blinds down, lights on' situation. The light shelf has a corrugated reflective surface designed to maximize high-altitude skylight reflection but to reject low-altitude, direct shortwave solar radiation. This almost doubles daylight levels in north-facing rooms, where adjacent buildings obstruct a clear view of the sky. Coupled with a barrel vaulted ceiling and light-coloured, fair-faced room surfaces, the light shelves help to produce an even, well-lit room environment.

The sealed façade might imply that occupants have minimum control over their environment. This is not the case, since users have manual trim control over the air supply volume, radiator output, window blinds and the luminaires, including daylight dimming override.

Conclusions

This is a building which raises the concept of the active façade to new heights, working in close conjunction with the room thermal capacity. It has many elements serving a wide variety of functions at different levels and for alternative orientations. The objective of a predominantly passive system has been realized, the only moving elements being the blinds operated by room occupants.

Traditional wisdom dictates that a building sealed to exclude external pollution will require air conditioning, with an energy consumption at least twice that of a naturally-ventilated building. In the case of the new Parliamentary Building, a system which integrates the building environmental approach with simplified engineering has the potential to use considerably less energy than even the naturally-ventilated norm.

This may well prove to be a landmark building, introducing a new generation of buildings which use their fabric to capture and repeatedly recycyle ambient energy, and use the same energy sources for the system's motive force. When incorporated into long-life buildings, this design principle is a major step towards zero-energy buildings, which will be fully realized when there is a major improvement in the efficiency of photovoltaic cells and a reduction in their price.

Summary

The buildings featured in these case studies have made use of a wide range of techniques, based on the sometimes varied views of their design teams. However, several themes recur, which should not be considered mandatory in all circumstances, but should be considered in appropriate situations:

- paying careful attention to window/glazing technology, in terms of light admission and solar heat gain, often making use of special shading techniques;
- using atria for a variety of purposes, with the aim of improving the internal environment;
- employing thermal mass effects to reduce temperature fluctuations and the need for air conditioning;
- avoiding the use of air conditioning by employing alternatives such as natural ventilation or mixed-mode systems where possible;
- using sophisticated control systems linked to novel technologies to optimize the indoor climate.

Recommendations and check lists

Energy-efficient design of dwellings

The regulatory regime

Regulation as a means of saving energy in buildings first developed in many countries as a result of the oil price increases dating from the crises in the 1970s. The prime intention was to achieve security of supply by reducing demand. However, during the 1980s and 1990s, this has been overtaken by the increasing awareness that reducing levels of carbon dioxide in the atmosphere must be the overriding motive for driving up insulation standards. Countries have responded to this situation with varying degrees of commitment.

Five factors influence the level of thermal efficiency imposed by regulation on new buildings:

- climate;
- cost-effectiveness;
- international agreements regarding the abatement of carbon emissions;
- the capabilities of the building industry;
- the attitude of the market.

Climate

In the past, it has been assumed that climate is one of the fixed factors in the energy use equation. This is now being questioned, not least in the UK. One of the predicted consequences of global warming is that climates will become much less stable. The UK, being positioned between two major weather-making features – the Atlantic and the Continental land mass – may well be a particular victim of this change. Weather patterns over the last decade add weight to the prediction.

The second factor affecting the UK is that the Gulf Stream is showing signs of weakening, as discussed in the Annex. The Gulf Stream has a moderating influence on the UK's climate; without it, the climate might be similar to that of Labrador, which is on the same latitude.

The climate data used to justify regulations are often crude averages, which fail to acknowledge considerable differences in climate between extremes within a country. In some states, like Sweden, the regulations take into account the worst climate conditions; in others, like the UK, the regulations are based on the more moderate regional climates of the country.

The justification stated by the DoE for the discrepancy between UK regulations and those of other north European countries is that the more moderate British climate makes high insulation levels unnecessary. As part of an investigation into the DoE, the National Audit Office commissioned the authors to assess whether this claim was justified. The research involved in this assessment (National Audit Office, 1994) included:

- selecting three countries which would be appropriate for comparison with England and Wales (Scotland and Northern Ireland have separate regulations);
- comparing the climate data;
- collating and interpreting the four sets of regulations and the 'deemed to satisfy' conditions therein;
- conducting a comparative analysis of energy consumption between the countries, using a standard dwelling located in Sheffield's climate, using both UK and Danish computer programs for the calculation.

The countries involved in the study were: England and Wales, Holland, Denmark, and Sweden. The regulations were those in force at the time of the study in 1993.

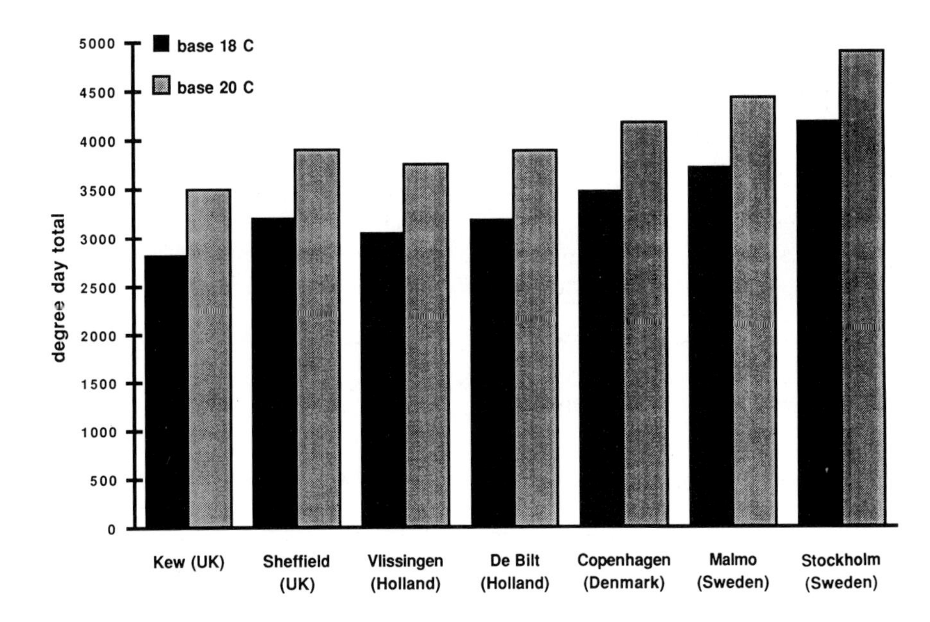

Figure 5.1 Degree day totals to different base temperatures at sites in the sample countries

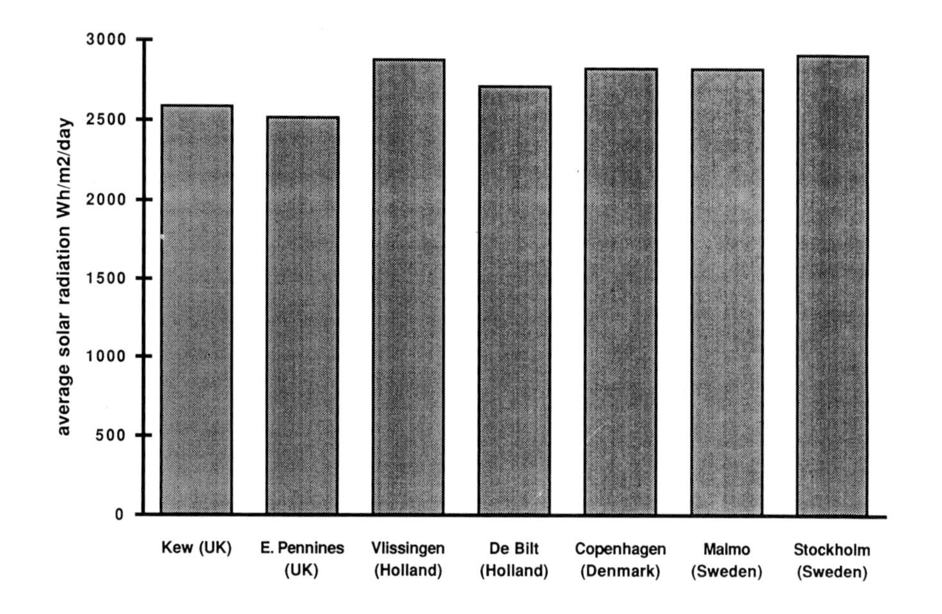

Figure 5.2 Average daily solar radiation at sites in the sample countries

Figure 5.3 Predicted space-heating energy consumption in the sample countries

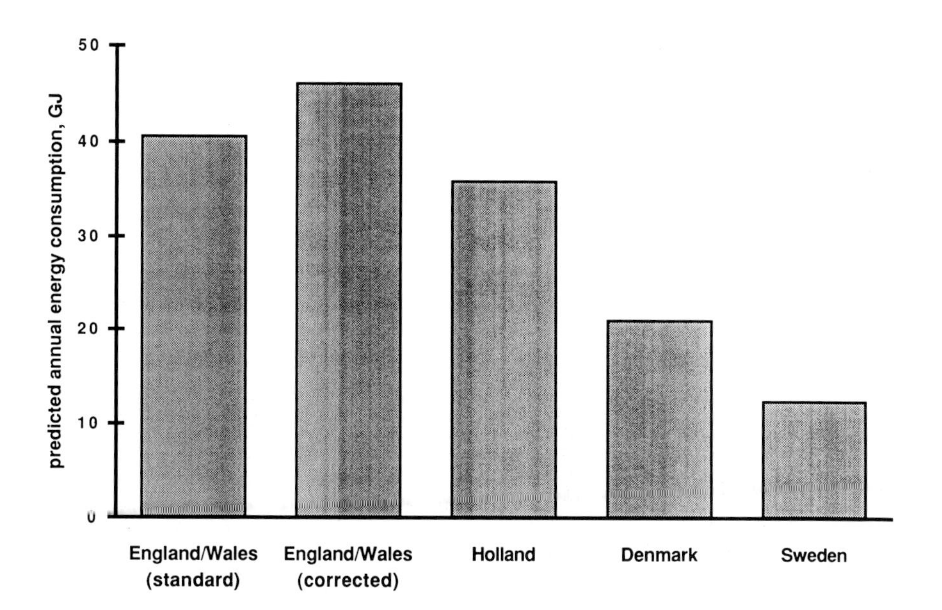

Climate comparison

The most useful, if simple, measure of climate for assessing heating requirements is 'degree days'. These represent the cumulative deficit or shortfall between outside air temperature and a base internal temperature. When the external temperature is below the internal base value, it is assumed that heating is necessary. The accumulated deficit is usually calculated on a month-by-month basis to remove the vagaries of day-to-day fluctuations.

The study took two base temperatures, 18°C and 20°C. It had been argued that the base temperatures for analysis should be lower in the case of England and Wales, because average temperatures in dwellings were lower than in some other countries. The authors contend that this is the result of the difficulties occupants experience in obtaining adequate warmth in dwellings in England and Wales, either because of poor design or low income – usually both. They do not subscribe to the view that people in England and Wales *prefer* cooler conditions than other European nations. There is good evidence that when energy-efficient homes are provided, the occupants first seek higher comfort temperatures, and only when these are achieved are cost savings realised.

The East Pennines region, in which Sheffield falls, was chosen to represent England and Wales since its climate is within 2 per cent of the weighted national average. The results showed that Sheffield is comparable to Holland and only slightly lower than Denmark in terms of degree day totals. The difference between Sheffield and Malmö (Sweden) was just over 500 degree days (see Figure 5.1)

However, England and Wales have lower annual averages of daily available solar energy, which means that they have less opportunity to reduce space heating energy by exploiting solar gains (see Figure 5.2)

The authors concluded that whilst there are differences in climate, their order of magnitude did not justify the discrepancies in building regulation energy-efficiency standards between England and Wales and the other countries. To underline this, the authors carried out an investigation which discounted the climate factor altogether.

In order to eliminate all variables except thermal regulations, a 100 m² semi-detached house of a type common in the UK was chosen as the basis of the comparison. Its single-glazing comprised 15 per cent of the floor area. The energy assessment was based

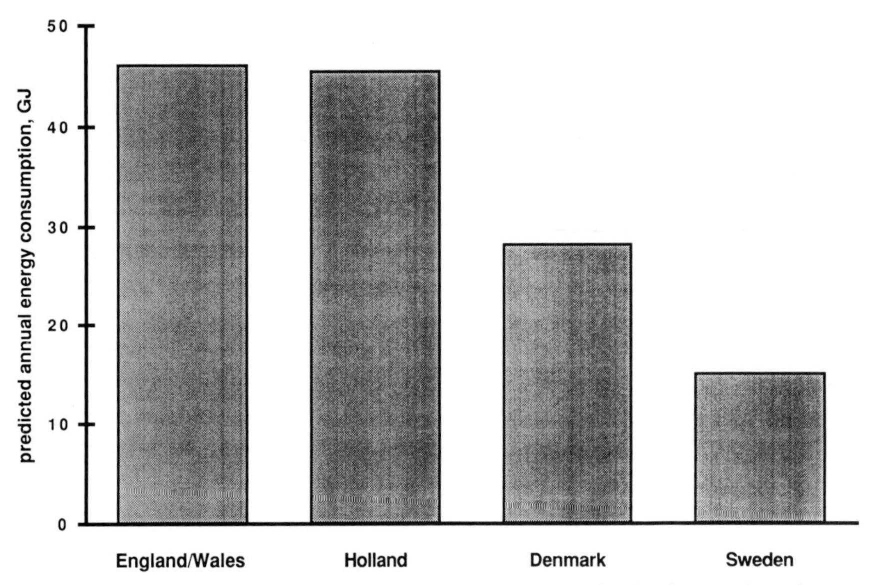

Figure 5.4 Predicted space-heating energy consumption in the sample countries, based on their own climate data

on a version of the Building Research Establishment Domestic Energy Model (BREDEM). The building regulations of all the countries were applied to the design, using the East Pennines/Sheffield area climate data in each case.

The results showed that the regulations for England and Wales led to a design with a space-heating load well over three times greater than a design complying with the Swedish regulations, and twice as much as a design complying with the Danish regulations. The predicted energy use in Holland was closer to that of England and Wales, mainly due to the fact that its regulations permit a higher ratio of window to floor area (see Figure 5.3). If a more accurate Danish calculation method is used to assess U values, the Swedish regulations are over four times more energy-efficient than those that apply in England and Wales.

Perhaps the most conclusive evidence of the poor performance of the English/Wales regulations was provided when a Scandinavian energy prediction tool was used to assess the energy use of dwellings in their own climate and with their own regulations, with corrected U values. One might have expected that energy use would be broadly similar in each case, but the analysis shows that this is not the case (see Figure 5.4). If all calculations were based on the same base temperature, the Danish dwelling used only two-thirds energy of the England/Wales house, whilst the Swedish example used just over one third (35 per cent).

Recent changes to the U value calculation technique and minor upgrading of insulation standards have modestly improved the performance of the England/Wales regulations, but they still lag far behind those of a number of European neighbours.

Cost-effectiveness

When devising amendments to the thermal efficiency section of *The Building Regulations*, Part L (DoE, 1995), the Department of the Environment was obliged to apply normal market criteria for cost-effectiveness. This meant imposing a discount rate of 6 per cent, which was said to produce a 13-year payback period. In fact, this would produce a return on investment which would outshine many other investment opportunities.

Other countries take a longer-term view, taking account of the probable pressure to include the cost of externalities in the price of energy. Even a 1 per cent reduction in the discount rate can make a significant difference. For example, whereas 40 mm thick wall insulation is deemed cost-effective in England and Wales, German cost studies based on a 5 per cent return on capital reveal that the most economical thickness of masonry wall insulation is 150 mm.

If regulatory authorities were to adopt the precautionary principle in this context, they would err on the side of generosity and promote greater use of insulation, on the grounds that the economic penalty for installing inadequate insulation would be much greater than for installing excessive insulation. If such sensitivity studies were to include carbon tax predictions, then the arguments in favour of high insulation would be irrefutable. However, the present UK Building Act makes it impossible to use economic criteria which include risks due to global warming.

Short-termism infects every aspect of cost-benefit analysis and fiscal policy. There needs to be a fundamental shift in attitude, both on the part of governments and the general public, prompted by an appreciation of the benefits of energy conservation and the negative consequences of inaction.

International agreements

The agreement reached at Rio in 1992 to reduce and maintain carbon emissions at 1990 levels by the year 2000 is the only binding commitment entered into by the industrialized countries. Even this modest target has put pressure on legislators to set carbon abatement targets which they might otherwise have preferred not to implement. The UK claims to be on course to meet this target, largely thanks to industrial stagnation and the large-scale switch from coal- to gas-fired generation of electricity. However, the EU predicts that the UK will fall 3 per cent short of the target by 2000, and the

International Energy Association puts the shortfall at 9.5 per cent (IEA, 1996, p.6).

In 1995, the international conference held in Berlin sought to persuade these countries to set targets beyond 2000. The UK has agreed to cut emissions by a further 5–10 per cent by 2010. Some believe this will be achieved automatically by the natural increase in the ratio of new, better-insulated buildings to old stock, and the improvements in the efficiency of energy-using technologies. In contrast, Germany has undertaken to cut its carbon emissions by 25 per cent by 2010. This is already making its impact on the construction industry. Certain countries, such as Switzerland, are also unilaterally imposing a carbon tax, and Denmark has had such a tax for some time. The UK has a form of tax in the Non-Fossil Fuel Obligation (NFFO), which is a levy on electricity generation using fossil fuels. Of this levy, 98 per cent has gone to support nuclear power and build up insurance against nuclear power station decommissioning costs. There is pressure on the UK government to maintain this tax in a different form after its official deadline of 1998, in order to support the upgrading of the housing stock.

The capabilities of the building industry

Another argument which is used to justify thermal regulations that many regard as inadequate is the low level of competence within the industry to deal with change and new construction techniques. It is true that high insulation standards require high standards of competence, since relatively small construction faults have an increasing impact the more energy-efficient the design becomes. In the UK, much of the housebuilding activity is carried on by small builders, who, allegedly, do not have either the financial or intellectual capacity to adapt to serious changes in construction techniques. The National Housebuilders' Federation has tended to use its influence to oppose improving energy-efficiency standards by regulation. On the other hand, some of the large-volume housebuilders have been innovative in constructing demonstration low-energy schemes.

In contrast to the UK, when Sweden substantially upgraded its standards in SBN 80, it gave advance notice to the industry and subsidized training programmes so that the operatives were fully competent by the implementation date. The later SBN 89 brought all regulations into a single framework, and its changes did not require any radical adaptation to new technology.

To adopt inadequate thermal efficiency standards in response to a perceived low level of ability on the part of builders is to yield to a doctrine of despair. Time and again it has been shown that, when challenged, small builders can rise to the occasion – as, for example, in the case of the extraordinary Vales' house, which claims to be autonomous (see Case Study 3.4).

The attitude of the market

Housebuilders will argue, with some justification, that energy-efficiency is low on the list of priorities of house buyers – location, size, style and even fittings come higher on the list. The main reason for this is that energy prices are still artificially low, and in any case, the present Building Regulations ensure a moderate level of thermal efficiency. In the UK, thermal efficiency has been driven by regulation.

In contrast, until recently, in Scandinavian countries it was market pressure which forced up standards. This is why there was no cry of outrage from the Swedish industry when the U value for external walls in housing was improved to 0.17 W/m^2°C, whereas builders in the UK successfully prevented the government from changing the equivalent value of 0.45 W/m^2°C for walls. It should be remembered that if Scandinavian calculation methods for U values were used, the UK wall U value would be nearer 0.6 W/m^2°C.

Conclusion

Achieving change in building design practices, especially during times of economic difficulties, will not be easy. Nevertheless, it is clear that, in some states, the national attitude enables governments to implement changes that significantly reduce the energy used in buildings.

It bears repeating that buildings account for the largest fraction of a developed nation's energy demand, and within the building sector, housing is the largest single component. If the energy-efficient design of dwellings cannot be achieved, the prospects for overall improvements in energy-efficiency are poor.

Maybe, given the free market ethos which prevails in many industrialised countries, a means of curbing carbon emissions will need to be implemented which places the onus on designers and building users. The 'carbon budget' system described in Chapter 9 may be one option.

To conclude this chapter, it may be useful to provide a summary of recommendations for realizing the goal of energy-efficient housing.

Checklist for energy-efficient design of dwellings

Minimizing heat loss

External features

- Provide shelter from prevailing winds in winter.
- Consider likely air flow patterns when designing site layout and door positions.
- Design the arrangement of building groups to reduce wind pressures.
- Non-solid shelter-belts (trees, shrubs, fences) with about 40 per cent permeability give the best protection.

Building features

- Compact building shapes reduce heat loss areas.
- The protection afforded by earth-berming and buffer spaces should be exploited, where feasible.
- Heated areas within the dwelling should be isolated from unheated spaces by providing insulation in the partitions between them.
- Glazing should be at least double, or triple-glazed low-E where possible.
- The areas of non-beneficial windows should be minimized.
- Window frames should be chosen to optimize energy-efficiency.
- Well-insulated windows allow more flexible heat-emitter (radiator) layout.
- The detailing of joints in the building fabric requires special attention.
- 'Cold bridges' must be avoided.
- Fabric insulation significantly above the minimum required by regulation is strongly recommended.
- Draught-sealing to reduce infiltration, in association with controlled ventilation, is important.
- Conservatories require careful design, being isolated from the main occupied area; at the same time, attention should be given to probable air flow patterns.

Solar heat gain

External

- The main façade of the dwelling should face close to south (± 30° approximately).
- Spacing between dwellings should be sufficient to avoid overshading.
- The benefits of sloping sites should be exploited whilst also guarding against their adverse effects.
- Areas with particular overheating risks should be considered when planning building layout and form.
- Deciduous trees and shrubs offer the advantage of summer shade.

Building features

- Internal layout should place rooms on appropriate sides of the building to benefit from or avoid solar heat gain.
- Westerly orientations for glazing, and in particular for windows into rooms with additional overheating risks (e.g. kitchens) should be minimized.
- Shading (external, if possible) for windows posing overheating risks should be provided.
- The effect of window frames and glazing bars on heat gain needs to be noted.
- Solar heat gain in conjunction with daylighting design must be taken into account.
- South-facing windows should be maximized, and north-facing windows minimized.
- High-thermal mass construction should be used for internal surfaces into which solar heat gain can be absorbed.
- Heat-absorbing floor areas should not be covered by carpeting.
- Conservatories or other buffer spaces can be used to pre-heat incoming ventilation air.

Systems

- The choice of fuel should minimize environmental impact.
- District heating or combined heat and power should be used where available.
- High-efficiency heating systems, for example, condensing boilers, are desirable.
- Space heating and hot water systems should be appropriately sized.
- Heating systems should be zoned and/or use thermostatic radiator valves.
- Controls, programmers, and thermostats should be appropriate, and their operation should be easily understood by occupants.
- The heating system should be appropriate to the thermal response of the building fabric and the occupancy schedule of the dwelling.
- Hot water storage vessels and distribution systems should be fully insulated.
- Heat-recovery ventilation systems are essential where effective draught exclusion measures have been implemented.
- Air infiltration/ventilation need to be controlled during winter.
- Ventilation of utility areas, bathrooms and kitchens should be considered to overcome moisture and condensation risks.
- Ventilation of hot air in summer may be considered.
- Installing heating in conservatories should be avoided in most circumstances.

The precautionary principle

The previous section dealt with dwellings and their energy consumption. A different scale of issues and problems applies to larger buildings, which are the focus of this chapter. Implementing changes in this field may be an easier task, since the natural conservatism which permeates the design and construction of dwellings is not so pervasive, and different issues are prominent. One major issue is the confidence which can be placed in the predictions of energy-efficiency and robustness in operation of novel environmental technologies.

In Chapter 4, a number of buildings were examined which illustrated the array of environmental and energy-efficiency measures which are being adopted in commercial and institutional buildings. At this point, it is worth considering how enthusiasm needs to be tempered with caution. Inevitably, there is a sense in which every building which employs new technologies is experimental. Some clients, like De Montford University, Leicester, UK, are very willing to accept this fact and allow their buildings to be used as a test bed, as in the case of their Queen's Building. This incorporated a number of different design approaches appropriate to environmentally sensitive architecture. Other clients are not so tolerant, and this has inevitably led to disputes. So why do things go wrong?

There is evidence that many recent offices designed to be energy-efficient are not performing as well as expected, sometimes by a margin of 25 per cent. This chapter will summarize some of the reasons for this. A factor which is often overlooked is that energy is comparatively cheap, and accounts for only 1–2 per cent of total occupancy costs, including salaries. This means that there is little incentive to incur extra expenditure to modify systems to meet the original performance specification. But this is in the context of the heavily-subsidized energy prices mentioned earlier. The picture changes when energy costs are related to profits, where they can range from 10–20 per cent of the total.

Where an environmentally effective building really does score is in the sphere of staff well-being. For example, a joint examination by the US Department of Energy and the Rocky Mountain Institute, California, of a number of refurbished offices found that renovations involving lighting and ventilation led to significant increases in productivity. A mere 1 per cent increase in productivity paid for a typical company's annual energy bill (Romm and Browning, 1995).

In a specific example, Romm and Browning state that when Lockheed commissioned a new 60,000 m^2 office complex, their architect persuaded them to invest in an extra 4 per cent in order to benefit from energy-efficient design. The result was that absenteeism dropped 15 per cent compared to their previous headquarters. Energy savings were worth $500,000 per year.

In a survey conducted in 1991, the Royal Institute of British Architects found that when architects recommended to their clients that they include extra energy-efficiency measures in designs, the results were:

- 32 per cent accepted the recommendations in full;
- 21 per cent accepted some of the measures;
- 47 per cent would not accept the extra initial cost, even though the longer-term benefits were explained. (RIBA Journal, February 1992, p.16)

The percentage of clients who accept the logic of energy-efficiency has improved since then, but there is still a significant section of the client community which is unconvinced by the arguments. They regard the potential extra 5–7 per cent cost as unacceptably high, especially if the building is to be let or sold on. Nevertheless, if the economics of the energy-efficiency on-costs are presented as a composite of capital costs offset by revenue savings, then the advantages clearly outweigh the disbenefits. Even in terms of capital cost alone, energy-efficiency can make savings. For example, high-frequency lighting, good reflecting luminaires and infra-red controls can save money because fewer fittings are required and less heat is produced, leading to a reduced cooling load. At the same time, there is an opportunity to install fewer switch drops, reducing cabling and simplifying fitting-out. This lighting strategy may also reduce the contract period, with obvious benefits in terms of an earlier occupancy date.

Not all the fault lies with clients. Many professionals are reluctant to negotiate new design territory, either because they fear falling victim to untried technologies, or because they will not make the effort to learn new construction techniques. All construction professionals operate in the shadow of 'professional indemnity', which can make them overcautious, and means they tend not to ask questions after completion.

Whilst design professionals are urged to work as a team, this is often difficult in practice. One factor which militates against integrated design procedures is fee competition, which sometimes reduces the returns for design work to less than cost. In a cut-throat world, designers across all disciplines are more often competitors than collaborators. One consequence of this is that services designers are often brought into projects at a late stage. Furthermore, a fee structure which is based on contract or subcontract cost operates as a disincentive to services engineers. They are less likely to embrace low-energy designs which do not rely on engineering hardware, as the costs of such hardware enhance their fees.

High-profile/low-profile

In the drive to reduce energy consumption, attention has tended to focus on the high-profile factors, such as insulation standards and heating/cooling installations.

Now that building regulations are driving up insulation standards, other factors are becoming more significant in terms of energy consumption, such as duct sizes and fan motors. In many, if not most, cases, fan motors are substantially oversized, leading to significant excess energy costs. Leaving computers switched on unnecessarily not only wastes energy directly, it also adds to the cooling load of the building. Lights are another source of concern which will be considered in more detail in Chapter 7. In many instances, substantial improvements to energy-efficiency can be achieved by paying sufficient attention to these low-profile details of design.

The 'high-tech demand'

Some designers are seduced by the imagery of advanced technology and install hardware that greatly exceeds the real demands of the building and its occupants. Designers should aim to avoid the technological fix and install only essential technology that is efficient, not overcomplex, and easy to use and maintain. Overcomplex systems which require elaborate maintenance tend to deteriorate fairly rapidly because service managers cannot meet the demands of the technology. In extreme cases, the system may be abandoned altogether.

One problem facing clients is the relative scarcity of information for non-specialists. Even the professionals have problems in this respect. The Energy Efficiency Office's Energy Consumption Guide 19 (BRECSU, 1991) is still a definitive document for the UK.

Operational difficulties

It is unfortunately the case that guidance/instruction manuals are often poorly-written and inadequate in terms of information. There is a universal problem with instruction manuals, because they are usually written by experts on the system in question who find it impossible to empathize with the uninitiated installer and operator. They fail to grasp the breadth of the knowledge gap. This problem seems to be especially acute in terms of services technology.

Another problem which is all too common is that installers are expected to comply with almost impossibly short completion deadlines. Commissioning is hurried to avoid activating penalty clauses in the contract. It may be less of a financial risk to commission the system properly after practical completion. If the system goes into operation at a substandard level of efficiency due to time constraints, service managers and office staff are at a disadvantage from the start. This is a recipe for high energy consumption and less than perfect comfort conditions.

Building-related illness

Over recent years, there has been increasing awareness of the phenomenon 'sick building syndrome', more accurately termed 'building-induced sickness'. Factors such as gaseous emissions from plastics in furnishings and fittings, or the frequency of fluorescent lights, have been implicated; poorly-designed heating and ventilation systems have also been identified as culprits, aside from the most spectacular problem of Legionnaire's disease. There have been numerous horror stories of badly-maintained systems providing a comfortable habitat for all manner of unmentionable life forms, as well as closed systems recycling bacteria and viruses, resulting in high levels of absenteeism.

Recent studies have suggested that sick building syndrome is also related to job satisfaction. Job satisfaction is more easily achieved in a pleasant, comfortable environment in which the occupants are permitted some degree of control over their surroundings. When energy-efficient design sets a good baseline of environmental conditions, the effort necessary to fine-tune comfort to an individual's personal preferences is reduced.

Inherent inefficiencies

A system designed to be energy-efficient can be totally undermined if the whole system has to be operated to meet a small demand. For example, in small-scale buildings, it is not unusual to find an entire heating plant being run in summer to supply hot tapwater. Unreasonable overcapacity is another problem. An elaborate and expensive chiller may be installed to meet the cooling demand of a few days per year or to supply cool air to a small number of prestigious rooms. The scale of such inefficiencies may go unnoticed because of the absence of proper monitoring systems. System efficiency can drop dramatically without management being aware of the problem. Often, a catastrophic failure is the first indication that something is wrong.

Sophisticated controls and electronic management systems combined with zonal submetering will ensure that faults are pin-pointed and system inefficiencies identified. The relatively small capital costs involved in such equipment will quickly be repaid. The operation of such systems must be supplemented by adequate supervisory and analytical input from knowledgeable staff.

Recommended principles for new office design

The following checklist identifies the main items which need to be considered in optimizing design:

- The first task is to persuade the clients of the benefits of environmentally sound and energy-efficient design. There is now convincing evidence that 'green buildings pay'.
- It is important that all members of the design team share a common goal and, if possible, have a proven track record in achieving that goal. From the earliest outline proposals through to construction

and installation, the design process should be a collaborative effort. Integrated design principles should be the rule from the first encounter with a client.

- The primary aim should be to maximize passive systems, and to reduce the reliance on active systems which use energy.
- Costs should be calculated in a composite manner, so that capital and revenue costs are considered as a single accountancy feature. This will help convince clients that the extra capital expenditure is cost-effective, even for buildings to be let or sold on.
- The claims made for advanced technology do not always match performance. It is important to select appropriate technology which achieves the best balance between energy-efficiency, occupant comfort and ease of operation and maintenance. At the same time, the best compromise should be reached between optimum performance and the requirements for the majority of the year. To provide significantly greater capacity for just a few days per year is not best practice.
- Lighting requirements should be clearly assessed to discriminate between general lighting and that required at desktop level.
- Clients should be required to explain in detail the nature of office routines, so that these can be properly matched to operational programmes.
- Building managers should be selected for their ability to cope with the complexities of the chosen building management system.
- Appropriate monitoring is necessary to enable day-to-day assessment of how systems are performing. Installing submeters, hours-run recorders and so on gives valuable returns for a small cost. Energy costs should be identified with specific cost centres.

Precautionary checklist

Common architectural problems

- The adverse effects of too much glass are often underestimated. Maximizing daylighting can produce problems for VDU operators.
- Inappropriate window design may lack refinement and ease of control.
- Controls and user interfaces may be poorly-designed.
- Fitting-out may contradict the original design intentions, leading to poor performance.
- There may be a tendency to highlight the positive and play down the negative. Downside risks may not be given the same weight as upside visions.

Common engineering problems

- Inappropriate standards regarding climate control, lighting and distribution of services may be adopted.
- Optimized engineering solutions may not be robust and flexible.
- Blind faith in technology tends to underrate the human factor, and fails to focus on the finely-tuned needs of occupiers.
- Mechanical ventilation may be inappropriately designed in terms of rate of ventilation, efficiency, operating hours and zoning. Special problems can occur with night-time ventilation.

Issues related to air conditioning:

- Whilst the avoidance of air conditioning, and thus reduction of plant, leads to a lower cost per square metre, it may result in a reduction in the proportion of the site which can be developed, because of the siting constraints posed by orientation,overshading, natural ventilation and perimeter area design considerations.
- Lower energy consumption is probable when air conditioning is not installed, but is not always easy to quantify.
- Lower running costs for the building when avoiding air conditioning may be gained at the expense of staff satisfaction. The alternative of a balanced natural/mechanical system requires sophisticated design techniques which are not always available.
- Natural ventilation is claimed to be more adaptable, but there is sometimes a failure to appreciate how this adaptability can be achieved.
- Naturally-ventilated buildings are claimed to offer greater occupant satisfaction. This can create variable climate conditions at any given time, and the level of occupant satisfaction is difficult to measure on a constant basis.
- The drive towards green design has encouraged the use of untried and inadequately-researched alternatives to air conditioning.
- Designers can place too much faith in arrows showing expected air flow when natural ventilation is used – a more rigorous approach is necessary.
- Natural ventilation is less controllable.
- There is less plant to maintain without air conditioning, but what there is may be more complex because of the modes of operation.

Common failures leading to energy wastage

- Designers tend to err on the side of caution. There is less risk in overdesigning than underdesigning. As a result, systems are often overpowered, and therefore wasteful.
- Often, the most convenient operating strategy is for switches to default to on, whereas manual-on and automatic-off is the more energy-efficient option.
- Small demands like domestic hot water can require whole systems to be in operation.
- Inadequate monitoring systems will fail to identify progressive failure.
- There may be intrinsic hidden faults in the system which impair energy-efficiency without any detectable effect on service.

The human factor

- People are more tolerant of conditions in a naturally-ventilated building than in sealed, air-conditioned boxes. For example, the acceptable range of temperature is wider, and perceptible air movement is more acceptable. The main reason is that people like to feel in control, hence the need to avoid excessive automatic control.
- Occupants generally do not have the patience to keep fine-tuning their building environment, and will tend to do what is most convenient. Robust, clearly-articulated systems are the answer.
- As a general rule, people find it easier to switch systems on than off, hence manual-on, automatic-off should be the norm. At the same time, the inertia factor tends to increase when people are in groups – there is a reluctance to be conspicuous, risking criticism.
- Sudden changes are disruptive, therefore automatic climate modifications should occur imperceptibly wherever possible.
- Awareness of the outside world is an important component of contentment. Most of the time, external views are perceived at a non-conscious level, but psychological studies indicate that the mind can make a very full response to the visual milieu without reference to consciousness.

Summary of recommendations

- The temptation to opt for complex and 'heavy' engineering should be resisted in favour of 'gentle engineering', in which loads on the HVAC system are kept to a minimum by appropriate, climate-sensitive building design.
- The passive potential of buildings should be fully exploited, and care should be taken to ensure that building form, controls, ventilation, blinds, windows and so on are all supportive of natural systems.
- Mixed-mode ventilation and cooling systems, with different services zoned according to use patterns and need, are often the most suitable strategy. Where natural and mechanical systems are designed to work in a symbiotic relationship, it is necessary to ensure that, in changeover conditions, the system does not default into concurrent operation of the mechanical and natural systems.
- Monitoring is essential to determine running costs and to identify critical failure paths before they lead to catastrophic failure.

Lighting: A special case

Current wisdom has it that office design should maximize natural lighting. One reason for this is that lighting is often the largest single item of energy cost, particularly in open-plan offices. Another factor is that occupants tend to prefer natural light, especially since certain forms of artificial lighting have been implicated as the source of health problems.

Design studies suggest that considerable energy savings can be made by maximizing natural light, particularly if it is linked to automatic controls. Passive solar studies claim that installing efficient and well-controlled lighting would reduce energy/carbon costs more than any other single item.

Recent analysis has thrown doubt on these assumptions (Bordass et al., 1991). Changes in office design and work routines have led to a reappraisal of the maximization philosophy. In addition, a growing body of evidence from user appraisal has shown that, in many cases, the claimed benefits of maximizing natural lighting have turned into clear disbenefits. As a result, recent occupancy studies have shown that artificial lights are used much more than predicted. There are many reasons for this, and this chapter will review some of the most prominent.

When the original research into alternatives to the permanently artificially-lit office space was carried out, work in offices was largely paper-based. At the same time, research and guidance in the past has been simplistic and has failed to focus on the real contexts in which people make decisions. For example, it is possible for a single decision by an individual to put a whole system into an energy-wasting state. Insufficient consideration is given to the fact that anomalous situations are often difficult to correct and it is easier to adopt the 'inertia solution'.

Nowadays, computers are the universal office tool, and excessive daylight can be a severe nuisance due to reflection from VDU screens. If lighting controls are not tuned to each individual workstation, this can result in greater energy use than in a conventional office. As an example, it has been found that all the lights in an office may be on because one person has drawn the blinds to avoid glare. Even where lights are zoned according to daylight penetration, these often do not relate to workstations, with the result that lights are left on all day to compensate.

One lesson which is being gradually learnt is that individuals will always select the least-cost option in terms of effort. It is not that people are inherently lazy, but that they will tend to resent expending effort on activities which they regard as the responsibility of management. For example, it is often easier to switch on lights than adjust blinds, and that is what happens when natural light levels fluctuate. The common 'inertia response' is to close the blinds and switch on the lights. Where daylight results in glare, individuals will adjust the blinds and artificial lighting to avoid discomfort and achieve an even distribution of lighting regardless of energy consumption.

In cellular offices, individuals take more responsibility for adjusting their light levels and optimizing the relationship between artificial and natural lighting. In open-plan situations, where no individual is responsible, blinds tend to be left closed if that was their position on the previous day, regardless of external conditions.

Photoelectric control

Where lights are operated electronically according to natural light levels, the systems can be either closed- or open-loop. A closed-loop system controls the lighting to top-up the daylight to achieve a given minimum illuminance level. Open systems measure external incident daylight to judge when to dim lights, but with no feedback regarding the actual levels of realized illuminance.

Blinds can override the controls of both open- and closed-loop systems. Complaints about the lack of finesse of such systems can result in management abandoning photoelectric control altogether. Where sensors in closed-loop systems are near windows, it is not uncommon to find occupants closing the blinds to activate the lights. Furthermore, in many cases there are not enough sensors and lighting zones to take account of localized variations in daylight due to orientation or shading. As a rule of thumb, to avoid the need to make fine adjustments, lights should not go off until the illuminance level is about twice the design level.

Glare

Another rule of thumb which is observed by designers is: 'If you can't see the sky, the daylight level is inadequate.' The result is tall windows to give maximum daylight penetration. This carries the attendant risk of glare unless workstations are properly positioned in relation to the window. In most cases, this will mean orientating desks at right angles to the external wall, and the VDU viewing axis parallel to the window plane.

One option is to resort to automatic blinds. Occupiers sometimes complain that the spontaneous action of such blinds is distracting and denies individual choice. Local manual override is the preferred solution.

Recent developments in glass technology also offer answers to this problem. In Germany, a form of thermochromic glass has been marketed. It is clear until its surface temperature reaches 30°C, when it turns opaque, cutting out 70 per cent of solar radiation. Its use for normal windows is obviously limited, but it has potential as a material for external fixed sun visors, eliminating all mechanical complications. Electrochromic glazing technologies offer further levels of control which may present a variety of shading options.

Another problem that occurs is that lighting and blind controls are not co-ordinated. Lights tend to stay on regardless of the position of the blinds. The operation of externally-positioned blinds can be impaired by adverse weather conditions, especially high winds.

Dimming controls

Closed-loop systems are designed to provide a constant level of desktop illuminance. However, as the level of outside light fluctuates, so individuals may wish to vary desktop light levels to minimize contrast. On a bright day, a desk lit at a constant level would appear gloomy.

Occupancy sensing

There are obvious advantages to light switching which is responsive to a human presence, but even this technology is not without problems. The adjustment of the sensors is a matter of fine-tuning. For example, they may not be sufficiently sensitive to the movements of people engaged in high-concentration tasks; alternatively, they may be so sensitive that passers-by trigger the switch, causing distractions. In practice, it is often difficult to locate sensors to suit occupancy patterns and work requirements, especially in open-plan offices where workstations may frequently be relocated. As stated earlier, the ideal solution is to rely on manual-on, automatic-off switching.

Occupancy sensors achieve their optimum value in service areas and circulation spaces. These are areas which are frequently overlooked, yet they can use more energy pro rata than office spaces. One common fault is that the positioning of switches and sensors does not take account of the contribution of natural light. This is a particular fault in offices with atria. In some of the worst cases, activating lights in an office area can switch on all the lights along the exit route, and in extreme instances, throughout the whole circulation area. A balance should be struck between optimizing safety and the profligate use of energy.

Switches

A common failing is that switches are not positioned logically in terms of their relation to the light fittings and behaviour patterns of occupants. Locating switches remote from fittings leads to uncertainty regarding the status of the lights. The answer would be to include a red 'live' light in each switch.

Remote infra-red switching is an efficient and effortless system, provided the operation zone focuses down to the size of an individual workstation. Where switching is inconvenient, the tendency, once again, is for lights to be left on permanently.

System management

One major reason why certain high-profile energy-efficient buildings fail to meet expectations is deficiencies at the level of system management. It may be that system interfaces are not well understood by staff. Even service managers and suppliers are occasionally not as well informed as they should be. The problem is exacerbated if the original software source is no longer available.

System complexity is another problem. If services managers are not conversant with the intricacies of a system, they will tend to operate it at or near its maximum capacity, on the principle that overkill masks lower-order problems and forestalls criticism. Also, overcomplex systems discourage interference and adjustment, in case the outcome is worse and defies a remedy. Calling out specialists to make adjustments can be expensive, thus tempting managers to operate the system at a level below its design efficiency.

Complex systems can also be inflexible. The system's program may no longer serve the functions of the building. In extreme cases, this has led to the complete abandonment of the system.

In some cases, the blame for poor system performance lies with office managers who fail to inform the staff of the operational characteristics, cost and energy implications of the system. For example, in one instance, staff were not told that pressing a switch twice would turn on extra lights. The human factor is of prime importance. Good communication between management and staff can achieve satisfactory performance from a less than perfect system. Inadequate communication can undermine the virtues of the best possible system design.

Sometimes, office managers fail to address the more subtle needs of staff, gearing the system to crude averages, with the result that nobody is satisfied.

The increasing popularity of flexible working hours also presents difficulties. Light controls may have been designed for fixed working hours and set lunch times. This is another instance of where the complications and cost of modifying the system to respond to new work practices may be unacceptable, and therefore the system is abandoned. A case in point is where all lights in an office are operated from a central control desk which is only staffed from 9 a.m. to 5.30 p.m. Since staff are now able to work flexible hours, it means that those working after 5.30 p.m. are obliged to leave the lights burning all night.

Another potential source of conflict between design and operation is when a single-occupancy office reverts to multiple occupancy. It is usual for the principal tenant to have overall control of the service. Variable working patterns and conflicting needs often mean that lights are left on unnecessarily.

A relatively recent trend in the production of buildings is for the design to be separated from the fitting-out. The architect and services designer may produce an elegant, energy-efficient concept, but this may be totally vitiated if the fitting-out contractor has not been informed of the energy-saving features and consequent operational constraints of the design. Even worse is the situation where the subcontractor deliberately ignores the design objectives of the architect and engineer in order to keep costs down. The problem of discontinuity between design intention and fitting-out can be particularly acute in the case of refurbishments.

Air-conditioned offices

These present a specific set of problems for designers. The psychological effect of a space hermetically sealed from the outside world is to suggest an environment designed to overcome nature and be wholly distinct from it. As a result, the inhabitants tend to regard it as natural that all the services should be used fully all the time. In high-rent air-conditioned offices which include the services in the rent, there is the attitude: 'We've paid for it, let's use it.' If, in addition, the façades feature solar tinted glass, even more lighting may be used to compensate for the constantly gloomy outlook.

Some conditions for success

Open-plan installations which offer occupant satisfaction and energy-efficiency tend to satisfy four conditions:

- The design is straightforward and comprehensible, avoiding overcomplexity.
- They have intelligible local controls with clear user interfaces.
- The system is robust and reliable.
- There is responsive and intelligent office and services management.

It is still comparatively rare to encounter an open-plan office which achieves low energy consumption for lighting combined with high daylight use, while producing a high level of occupant satisfaction. Those that do so display the following characteristics:

- There is an assertive client, who has formulated the system's requirements clearly and insisted from the outset on effective lighting controls.
- Following commissioning, there is intelligent management of the system, combined with responsive management at office level.
- Glare is reduced to a minimum by means of light shelves, overhangs, splayed reveals and deeply-recessed windows.
- The interior layouts position most desks at right angles to windows.
- Control systems are capable of responding to individual requirements with good switching design (often infra-red), and controls which are user-friendly and take account both of daylight levels and workstation layout.
- There is efficient lighting throughout the office, with high-frequency control gear and good optics.
- Design luminance has been set to achieve about 400 lux, with lower levels in circulation areas.
- There is variety in lighting, but without excessive contrast or the oppressive feel generated by installations with 100 per cent Category 1 luminaires.
- There are good levels of daylight in perimeter work areas, without causing excessive glare in interior spaces.
- Blinds are easy to operate, with a good range of adjustment, and need to be fully closed only in exceptional circumstances.
- Interior fittings and furnishings are light in colour and tonal value, with tall fittings kept to a minimum.
- VDUs can be moved easily to avoid glare.
- Circulation lighting is low-energy and well-planned and controlled, with full account taken of contributions from daylight. An important added advantage in corridors, etc., is the occasional external view.
- Daylight is maximized within circulation spaces, which has the effect of decreasing the use of artificial lighting in the office areas.
- The system has a high degree of inherent flexibility so that it can be fine-tuned and retuned to user needs.

Conclusions

A number of factors nowadays tend to direct designers away from fully-automated systems:

- Preferred conditions in offices are now complex and unpredictable, making it impossible to design for average needs. In the past, 'averaging out' was often an euphemism for a 'lowest common denominator' solution. Even so, designers and modellers are still reluctant to abandon their faith in fully-automated controls. Too often, there is still no clear analysis of what controls can achieve in practice, and how proficient people will be at operating and servicing them.
- Overcomplex systems can generate unpredicted consequences, and even episodes of total failure. In such instances, it is often perceived as easier to decommission the system than rectify it.
- Changes in office routine and design have revealed that opting for maximum daylight can produce irritating consequences, such as glare on VDU screens. Furthermore, maximum-daylight designs rely on the reliability and user-friendliness of blinds, and this reliance has often been misplaced.

The solution

The following list identifies some of the ways in which potential problems may be overcome:

- The aim should be to design straightforward, robust systems which are well within the abilities of service and office managers to understand and users to operate. This will deflect occupants from resorting to easy, energy-wasteful options.
- Achieving reasonable daylight levels is a lower-risk strategy than maximizing daylight. Large windows rely on an array of ancillary devices, with all their potential for malfunctions.
- Lighting systems should, as a general rule, be designed to be switched on manually and to default to off when not required.

Prospects for the future of energy-efficient design

Futurology is the most inexact of all the sciences. However, as computer models become increasingly sophisticated, so the foundation for prediction becomes firmer, and there are technical developments in progress which might reasonably be extrapolated. The short- to medium-term prospects may be considered under four headings:

- building-related energy production technologies;
- the efficient use of energy;
- developments in storage technology;
- the wider design implications of climate change.

Energy production developments for buildings

Considerable research endeavour is being directed towards raising the efficiency and lowering the cost of photovoltaic cells. At present, the peak energy conversion rate is about 17 per cent in ideal conditions. Costs are high, partly due to the fact that the technology still awaits the benefits of economies of scale. As stated earlier, Germany is stimulating the uptake by subsidizing the installation of PV systems in 2,000 homes, whilst Japan aims to have installed them in 70,000 homes by 2005.

Because of the considerable market potential of PV technology for the developing world, substantial research capital is available in certain countries. Therefore, it is reasonable to expect increases in efficiency over the next few years. This would make PV systems a highly attractive proposition for all buildings, despite present modest energy prices.

Developments in glass technology will have an impact on design. Earlier, reference was made to thermochromic glass developed by the Fraunhof Solar Institute in Freiburg, Germany (see pages 51-4). Its potential for providing shading to transparent insulation materials was mentioned, as was the possibility of employing it for solar shading. If thermochromic or electrochromic solar shades were integrated with PV cells, then the whole nature of the southerly façades of buildings could change. Part of the function would be electricity generation, and it is likely that on many days there would a surplus of electricity, which could either enter a storage system or be sold to the grid. For night-time and winter conditions, the grid would be used to make up the electricity deficit. The aim should be to achieve a surplus of supply over demand in relation to the grid, so that the net result is at least a zero-energy building.

Another technology which is receiving considerable research effort is the fuel cell. Improvements in efficiency combined with developments and economies of scale for PV cells would mean that PV hydrogen-powered fuel cells would be an economic proposition, especially for larger buildings. The Freiburg self-sufficent solar house (see Case Study 3.2) is involved in proving that this technology is feasible, even at the domestic scale. Within ten years, it could be commonplace.

A recent innovation by Pilkington plc is mirror glass in which the reflective surface is provided within the manufacturing process, and not applied as an additional coating. This means that curved solar collectors can be made at much lower cost than conventional mirrors. The economics of passive solar energy could be transformed by this development.

Some building complexes supplement their electricity supply with wind power. Even at the domestic scale, this has proved cost-effective where the cost of a grid connection is high. Projects have begun to appear which use wind power in less visually intrusive ways than the horizontal-axis windmill. One example is an office tower in which the circulation tower is detached from the main block, and the plan form curved to maximize the aerofoil effect. Small-bladed turbines are positioned at intervals in the space between the blocks. Derek Taylor of the Open University has produced a design for a single-bladed horizontal-axis turbine which can be situated along the ridge of the roof of a building. This also exploits the aerofoil section, but this time in the horizontal plane as a function of the roof. It is comparatively inconspicuous, and should not be intrusive in terms of noise.

Another system called aquifer storage makes use of the constant temperature of ground water. Two or more wells are sunk into the earth beneath a building. In summer, when there is a demand for cooling, the cold ground water within the wells at 6–10°C is used to cool a water circuit via a heat exchanger, which serves to cool the ventilation air. The warm water is then returned to a second well at a temperature of 15–20°C, where it is stored. In winter, this warmed water is pumped to a heat exchanger, thereby warming the ventilation air. The Netherlands has 19 such projects completed or under way, with an estimated annual primary energy saving of 1.5 million m^3 of natural gas equivalent.

Hydropower possibly shares with the horse the position of the oldest means employed by humans to supplement their own muscle power. In the early twentieth century, a further major advance was achieved by linking water-wheel technology to the generation of electricity. In the main, this was confined to major projects. However, as the energy picture changes, so small- to micro-scale hydropower becomes increasingly attractive.

Riverside buildings might be in a position to exploit small-scale hydropower. This could take three forms. The most straightforward is the 'run of river' system, which diverts part of the flow through a cross-flow turbine or a Kaplan turbine which has variable-angle blades. Another method is to dam a river to create a head of water. At present, the minimum efficient head of water is of the order of a few metres. However, turbine development may well make a head of as little as 2 m a practical economic proposition. Already, a system with a head of 2.2 m is in operation at Burgdorf near Berne,

Switzerland, with a capacity of 70 kW. In many instances, medieval mill races provide an opportunity to make use of existing works to provide power at relatively low capital cost. Where contours allow, diversion of water courses to give higher heads will permit larger schemes.

Most sites for large-scale hydropower generation in the developed world have already been exploited, where this is technically feasible. The main prospects for future developments lie in smaller-scale, local schemes.

Another option is to exploit tidal energy. Medieval tide mills, such as that in Woodbridge, Suffolk, captured water at high tide in a pound, releasing it when there was sufficient head to drive a water-wheel. This is still a feasible technology, which provides reliable, intermittent electricity.

All these technologies exist, but the present energy cost regime makes few of them cost-effective. It must be emphasized that, as and when fossil-based energy costs come to reflect the cost of the damage they cause to health and to the upper and lower atmosphere, then there will be a much stronger financial incentive to exploit zero-polluting technologies. Additional encouragement will be provided by improvements in energy-storage technology.

Energy-efficiency

If the emerging technologies for energy production, conversion and storage are to have an impact on global energy demand, it is vital that buildings use energy much more efficiently. The potentially dramatic changes necessary to reduce carbon dioxide emissions can only partly be addressed by substituting more environmentally benign, renewable energy sources. Much can be done to reduce energy use: the preceding chapters detail the wide range of options which can be taken up by building designers.

As an example, the effect of the prospect of carbon taxes on design attitudes can be seen in Switzerland. Legislation is being proposed which will penalize non-renewable energy sources while providing a subsidy for energy-efficiency gains. As a direct result of this, a 'solar city' is nearing completion at Plan-les-Ouates near Geneva. This major housing development will exploit three energy-saving technologies.

First, two-thirds of the roof area will be covered with black-coated stainless steel solar collectors to provide hot water. These are less efficient than the most advanced glazed systems, but they are much cheaper and can be installed rapidly to form an integral part of the roof. Their lower efficiency is more than outweighed by their economics and relative ease of installation. Most of the hot water will be stored in two tanks per block, each with a capacity of 50,000 litres. They will supply domestic hot water in summer, and supplement space heating in winter.

The second technology involves a double-flow ventilation system which will provide all apartments with 1–2 air changes per hour, with full heat-recovery. The extract air will provide ventilation for the car park.

The third feature of the project is an 'earth source pre-heater'. This consists of a 6 km grid of pipes in the earth beneath the car park. In winter, fresh air for ventilating the apartments will first be passed through the grid and warmed by up to about 10°C, due to the fact that, even here, the earth temperature never falls below that value. In summer, the system will run in reverse to provide space cooling.

Prospects for storage

Renewable energy is almost invariably intermittent. This inherent disadvantage can be overcome if it is coupled to an efficient storage system. Battery technology has not made the advances predicted a few decades ago. In the future, attention may focus on the range of storage and generating opportunities presented by hydrogen. Recently, some bizarre claims have been made for technologies which produce hydrogen over unity, that is, the output of power is greater than the input, thus undermining the second law of thermodynamics. Nevertheless, there are clearly prospects for the use of new technologies and techniques in the future.

For the moment, reliance must be placed on improvements in conventional electrolysis to produce hydrogen at an economic rate. The most acceptable method of storing hydrogen is to fix it in metal hydrides such as iron/titanium hydride; hydrogen is released as needed by the application of heat. The parallel developments in PV, electrolysis, hydride and fuel cell technologies will offer individual buildings the opportunity to become autonomous in terms of electricity demand.

The most promising hydrogen storage technology under development involves nanofibres of graphite. The gas is stored under pressure and it is claimed that a graphite pack could power a car for 8000 kilometres.

Another storage system which might well prove attractive because it makes small demands on space is flywheel technology. This is a technology pioneered in vehicles. However, its best potential lies in the storage of energy in much larger quantities and for considerably longer periods than is needed in vehicle technology. The ultimate efficiency will be realized with a levitating flywheel supported by magnetic fields produced by high-temperature superconducting ceramic technology. In Japan, researchers have developed such a flywheel, which is made to rotate by magnetic induction to a speed of 3,600 r.p.m. This represents a storage capacity of 10 kWh. If situated in a near-vacuum, the slow-down, and thus energy loss, over 24 hours is barely measurable. Energy is generated by permanent magnets in the disc inducing an electrical current in a coil. Satellite technology is pushing the potential of flywheel storage to new limits. In the USA a flywheel has been developed which can attain 600,000 rpm with an energy density of 250 Wh/kg.

High temperature super-conductivity will allow massive amounts of electricity to be stored in a ring of super-conducting cables with no power loss and tapped to meet demand.

Building energy storage

Electrical storage systems are relatively short-term facilities, mainly used to flatten the peaks and troughs of supply over demand. Much longer-term energy storage potential lies in accumulating the low-grade heat associated with building heating and cooling processes.

Systems exist which operate over a yearly cycle, offering seasonal storage. Summer heat is accumulated to provide background heat during the winter heating season. As yet, this relatively low-technology form of energy storage has not achieved acceptance, and experimental development is still under way. Fundamental changes in the energy price structure or carbon taxes will quickly change the perception of this robust and almost maintenance-free technology.

Whilst daily storage systems, such as overnight ice-store production for use in air conditioning applications, are being used in some instances, from the point of view of someone commissioning an office building, seasonal storage is not cost-effective at present, considering the amount of space which must be dedicated to such a system. Whatever the storage medium – bricks, water or phase-change materials – they need to occupy a significant volume if they are to sustain their output over an average heating season in temperate climates. However, if excavation is necessary for other reasons, such as cut-and-fill or poor ground conditions, then even at today's energy prices, it would be well worth considering.

In the medium to long term, the energy cost burden born by buildings could well become an appreciably larger percentage of overall annual costs. Buildings are now being designed to last, so there are likely to be radical changes in attitudes towards fossil fuels during their lifetime.

Other design implications of climate change

Past evidence has shown that climate changes can be sudden and catastrophic. This is because a point is reached when adaptation processes are stretched beyond their limit, and there is a step change, or catastrophe point, to adjust to a new level of stability. Information gleaned from ice cores has shown that such changes can occur well within the lifetime of an average building. This could mean that the original climatic design of the building would become completely inappropriate. How to design for such a contingency will be problematic and very much dependent on the quality of prediction from the IPCC Scientific Committee.

The increased intensity of convective currents will increase the frequency and intensity of storms, especially within the zone in which sea temperatures rise above 26°C. As explained in the Annex, this is the minimum temperature at which tropical storms are generated. The construction industry may not be alert to this danger, but this is not the case with the insurance industry, which is bracing itself for the day when a supercyclone hits a major city. Designers should now be considering this order of wind loading, not just in a straight line, but under cyclonic conditions. Buffeting can inflict much more damage than a constant, linear high wind.

Another prediction is that specific climate zones will experience a much greater range of temperatures. After one of its hottest ever summers in 1995, the UK experienced the tenth coldest December this century, and in 1996, the dullest January on record, and the coldest February for 20 years. This was followed by a succession of colder than average months up to the end of May. The term 'global warming' implies that temperatures will simply increase – this is not the case. In fact, the UK will experience much colder winters if the Gulf Stream changes direction, which is another possibility as a result of global warming. By the beginning of the next millennium, designers will probably have to revise their climate criteria, and the consequent servicing of buildings, to take account of greater extremes of temperature.

Changes in rainfall patterns will lead to droughts in some places and flooding in others. At the very least, this will have implications for the design of foundations.

Rising sea levels, mainly due to thermal expansion of sea water, will greatly increase the risk to buildings in coastal locations. In the UK, insurance companies are already showing reluctance to cover buildings near coasts below the 5 m contour. The prediction is that small rises in sea level will have a disproportionate effect on storm surges, increasing the risks to buildings along coasts and beside estuaries. In such locations, it might be prudent to ascribe low-grade use to the ground storey, and, in particular, to keep mechanical and electrical services out of reach of possible flooding.

One further possible outcome of global warming is an increased incidence of *tsumani* (tidal waves), caused by the eruption of undersea volcanoes. Changes in the course of ocean currents triggered by global warming are considered to be the likely cause of such increased activity. However, at this point, architects could be forgiven for regarding this as one danger too far, and simply opt to take precautions against normal seismic activity.

At the present rate of progress towards carbon abatement, it is probable that the IPCC recommended annual emission level will not be achieved until at least the second half of the 21st century. In this case, considerable climate damage and social disruption will have taken place in the mean time. Buildings, as the principle users of energy, will doubtless become the main targets in the war against climate damage. Not only will they have to shoulder responsibility for making the major contribution to carbon abatement, they will also have to adapt to the severe climatic consequences of global warming. Life for architects and engineers in the 21st century will not be easy.

Sustainability: Attainable or impossible?

Current issues

There are two factors which threaten the prospects for a sustainable future:

- the social disruption arising from global warming and its climatic consequences;
- the rising world population.

Both will make their impact on architecture as more and more people are packed into energy-profligate cities. At the moment, the urban population is approaching half the world's total; within 20 years, it will probably reach 60 per cent of the predicted world population. It is believed that at least 300 million people will populate 21 megacities of over 10 million inhabitants, none of which will be in the west. Tokyo is expected to peak at 28 million.

This escalation of urbanization will enormously increase demands on energy; in the developing countries, most of that energy will be derived from fossil sources, and will therefore be particularly carbon-intensive. There will be pressure on the West to make a determined effort to compensate for this steep increase by making dramatic reductions in its carbon emissions, most of which will still come from buildings despite the rise in transport-related carbon emissions. In 1993, the World Energy Council produced a report called *Energy for Tomorrow's World*, which predicted that in 25 years, energy use will have increased by 88 per cent compared to the 1990 level. More recent predictions regarding the growth rate of cities along the Pacific rim make even this prophecy look conservative.

Since the technology to create almost zero-energy buildings is already available, the goal for the future must be to design buildings which will exceed the zero target and make a net *contribution* to the electricity grid. It must also be accepted that this will not happen if there is no state intervention in the energy market – either through legislation controlling the design of buildings, or some form of tax on carbon.

Targeting the design of buildings has been the policy up to now. However, for buildings of the future, it is likely that this method of conserving energy will not achieve the targets that developing global warming will make imperative. The two main reasons for this are, first, that construction-related regulations only embrace new building, and second, that the wide range of energy-efficiency criteria in the developed countries suggests that to achieve a consistent curb on carbon emissions across all these countries by controlling building standards would prove much too complex. The best hope lies with targeting carbon.

A rigorous examination of the potential of a carbon tax to achieve the IPCC carbon reduction goal of 60 per cent of 1990 levels has been carried out at the Department of Applied Economics at Cambridge University. Dr Terry Barker et al. (1994) have concluded that, to be effective, a carbon tax will have to be set at around $100 per barrel oil equivalent (b.o.e.). The tax would have to reach this level by 2010, replacing current value added tax (VAT), to reduce carbon emissions to the recommended level by 2050.

Even this draconian level of tax assumes that, by 2040, most energy would come from renewable sources, and that drastic conservation measures would be in place. At the same time, the price of goods and services would have to reflect the real cost of their carbon-intensity. In other words, their price would incorporate the external costs, namely those arising from the monetized value of damage to both the lower and upper atmosphere.

Whereas current energy costs in commercial buildings represent only a small percentage of the total annual costs – as little as 1–2 per cent of total occupancy costs, including salaries – the impact of such a tax would dramatically alter this proportion. It would require catastrophes on a colossal scale to enable politicians to impose such taxes. One advocate of carbon taxes, Victor Anderson, has stated that monumental political will normally offered only by dictatorships might be the only way such a tax could be implemented (Anderson, 1993).

The carbon budget

One way to avoid the political fallout from such a tax would be to develop the principle of the 'carbon budget'. In the construction sector, this would constitute a lifecycle allowance of carbon for individual buildings. At this stage, this may appear simplistic, since much of the information needed to make the concept work is not available. This will certainly change, as extensive international work is in progress to establish the levels of energy embodied in materials. At present, this work is hampered by the reluctance of manufacturers to supply information which they believe might place them at a market disadvantage. In this case, legislation is necessary to make disclosure of information such as embodied energy levels compulsory.

Elements of the budget

Returning to the lifecycle budget, this would consist of five elements.

Embodied energy

Measuring embodied energy would involve assessment with reference to:

- extraction of the raw materials;
- the manufacturing process;
- transport involved in the manufacturing process;
- transport to site.

Credits

Credits to the budget would be offered for factors such as:

- using recycled materials;
- generating energy from renewable sources, such as landfill methane for brick manufacture;
- the extent to which the manufacturing process is energy-efficient – for example, steel manufacturers contributing to the local combined heat and power grid.

Construction

The energy involved in the construction process would also be taken into account, involving such factors as:

- the height of the building;
- special site conditions;
- the ratio of contract cost to length of contract.

Energy in use

Each building would have an allocation of carbon for a specified lifetime, which would be skewed in favour of long-life buildings. Factors determining the 'in-use' budget would include:

- the building type – domestic, commercial, institutional, industrial, etc.;
- type of use, including rate of occupancy;
- the floor area;
- the total volume;
- the climate zone;
- special microclimatic conditions.

Other issues

For each building type, there would be a statutory lifetime norm. Every year of life in excess of the norm would gain carbon credits for the owner/occupier. Conversely, if the building were to be demolished before its lifetime norm, this would incur a carbon penalty.

Demolition analysis should include:

- factors such as location and the immediate environs;
- the proportion of material capable of being recycled – either processed or unprocessed;
- the quantity of material constituting low-grade waste;
- the quantity of toxic waste;
- transport energy costs.

The lifetime budget

The building would be given a lifetime allocation of carbon, expressed in units of, say, kilogrammes of atmospheric carbon emission. The allocation would assume a gradual improvement in energy-efficiency over that period, combined with a progressively greater contribution from renewable energy sources. At the outset, the building user/owner would receive a chart of the allowed carbon expenditure for the four phases of the building. The end of each phase would be a carbon tax point. When the building was in use, quarterly/annual energy bills would indicate the carbon equivalent of the energy used. If not already known, it would quickly be realized that different fuels offer widely varying kilowatt hours for a single carbon unit, as can be seen in Figure 9.1.

The driving force behind such a scheme would be that for every carbon unit expended over the budget allowance, there would be a severe cost penalty. If the system was to be credible, then the over-budget carbon tax would have to be at the rate of about $200 b.o.e. per kilogram of carbon (kgC). The level at which that penalty was set would determine the effectiveness of the scheme, and at the same time would create the possibility of a market in carbon units. For example, a building owner might decide to commission a building with low embodied energy that was highly insulated and made maximum use of natural resources. As a result, the annual carbon balance would fall well within the recommended budget allocation, meaning that at the end of the building's life, there would be a substantial carbon credit. The consumer

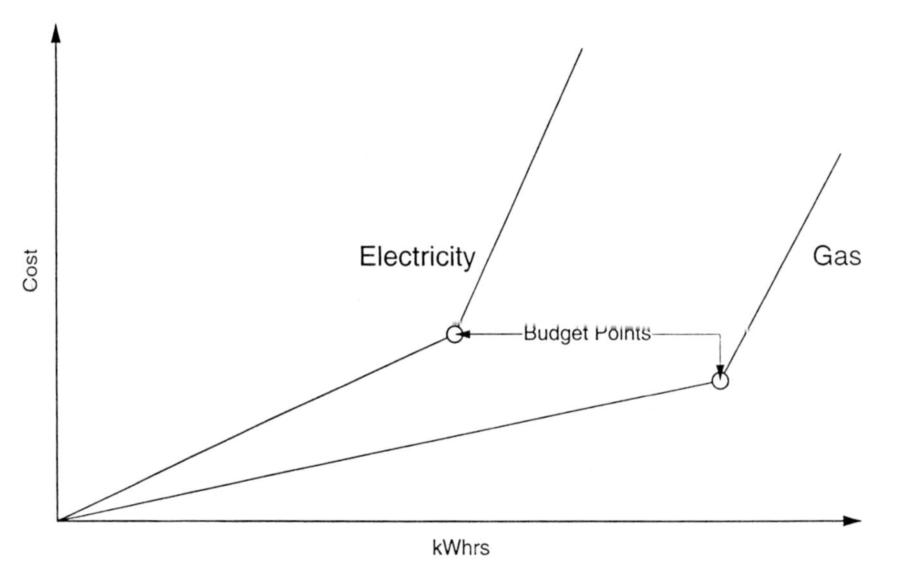

Figure 9.1 Application of the carbon budget

might choose to cash in carbon credits on an annual basis in anticipation of a terminal credit balance. If the penalty rate was $200 b.o.e./kgC, then the market value of carbon units should be in the region of $180 b.o.e./kgC.

In contrast, another building owner might choose to create an energy-inefficient building, balancing the cost savings of doing so against the expected carbon penalty. It might conceivably be cost-effective for that owner to buy extra carbon credits on an annual basis to make up the deficit. However, this would be a risky strategy, since the level of carbon penalty might well rise in the face of more urgent demands to cut carbon emissions.

The effect of this scheme for designers would be that the responsibility for the energy performance of the building would fall squarely on the shoulders of the design team. The person or group procuring a building would specify the energy performance *and* the lifecycle carbon budget for the building, based on the current carbon-intensity of the fuel mix. This means that the client(s) would have to be much more precise than is the norm at present regarding how the building would be used, and would have to take responsibility for the efficient operation of the building management system. At the same time, the design team would be responsible for advising the client if the brief and proposed use would be likely to lead to a building exceeding the projected lifecycle budget.

Perhaps the greatest value of such a scheme is that it would embrace existing buildings at the inception of the scheme. Naturally, such buildings would only be subject to the 'in-use' element of the budget spectrum. However, the same criteria would apply as for new buildings, thus encouraging the upgrading of their thermal efficiency.

There is no doubt that within this carbon-centred regime, the energy-efficiency of a new building would have a significant impact on its commercial value, whether for selling on or letting. Similarly, the state of the carbon budget for an existing building would have a major influence on resale or lease value.

This scenario has been developed to illustrate how architects and building owners may be affected as the international pressure to curb carbon emissions builds up in the face of increasing climate damage and social disruption. The relative apathy of the present time should not lead designers and procurers to be complacent. Most buildings being designed now could still be in use when such an energy regime came into operation.

The principle of budgeting for carbon could also be applied to manufacturing processes and transport.

Towards sustainability

A great deal is heard these days about 'sustainability', not least in the context of buildings. It is one of those words capable of a wide range of interpretations, which means that it can be more a barrier to progress than a facilitator. The Brundtland Commission defined it as leaving the planet to the next generation in no worse state than that in which the present generation found it (WCED, 1987, p.43). This is inadequate, since the planet inherited by the present generation was already in a condition of unsustainability.

There needs to be a definition which recognizes the scale of the threat facing future generations. According to David Pearce (1993, Chapter 2), there are two approaches to sustainability, which he calls 'weak' and 'strong'. Weak sustainability means that the aggregate of capital stock passing to the next generation remains constant. However, the depletion of natural capital can be compensated for by increased man-made assets, such as buildings, roads and machinery. 'Natural capital' or 'critical natural resources' includes soil, biodiversity, the hydrological and carbon cycles and the ozone layer.

Strong sustainability assumes that there are ecological assets that cannot be replaced by man-made artefacts. There is a need to move progressively towards preserving critical natural capital and living off the interest they can generate (Pearce, 1993).

Adoption of strong sustainablility will require a major shift in attitude, but it is the only chance for the planet in the face of the two relentless processes under way: population rise and global warming.

Those who design and refurbish buildings are in a position to spearhead this change in attitude, since buildings, as the largest energy-use sector, are also the most amenable to transformation. Not only should all those involved in building design and procurement aim at zero-fossil energy buildings, the ultimate objective should be zero-natural capital. Renewable materials such as timber and recycled materials, as well as renewable energy sources, should be the norm in the future.

Since it is the Western nations which caused the global warming problem in the first place, it is there that the sustainable building revolution should begin, and its technology should then be transferred to the developing countries which now pose the greatest threat to sustainability.

Annex: Global warming: The evidence and likely consequences

The long-term welfare of the planet, and particularly its human inhabitants, is now widely considered to be threatened by a range of factors, most of which are of human origin. They range from ozone depletion to the reduction of biodiversity, from erosion of the earth's capital assets to the stresses which will result from the predicted rise in world population.

The focus of concern is the earth's ability to support the diversity of life currently existing – its 'carrying capacity'. Much of the threat to long-term sustainability revolves round the rate at which humans are eroding the planet's capital assets.

However, the aim of this Annex is to consider the evidence supporting global warming and its climate consequences. In particular, the problems associated with the increase in levels of atmospheric carbon will be explored, starting with a brief explanation of the carbon cycle and its impact on global temperatures. This is done in the belief that biomorphic architecture, in the true sense, must be the result of conviction about the issues driving the future of the planet, and the importance of buildings in that process.

The carbon cycle and global warming

Carbon is the key element for life on earth. Compounds of the element form the basis of plants, animals and micro-organisms. Carbon compounds in the atmosphere play a major part in ensuring that the planet is warm enough to support its rich diversity of life.

The mechanism of the carbon cycle operates on the basis that the carbon locked in plants and animals is gradually released into the atmosphere after they die and decompose. This atmospheric carbon is then taken up by plants, which convert carbon dioxide (CO_2) into stems, trunks and leaves, etc., through photosynthesis. The carbon then enters the food chain as the plants are eaten by animals.

There is also a geochemical component to the cycle, mainly consisting of deep ocean water, estimated to contain 36 billion tonnes of carbon, and rocks, estimated to contain 75 million billion tonnes of carbon. Volcanic eruptions and the weathering of rocks release this carbon at a relatively slow rate.

Under natural conditions, the release of carbon into the atmosphere is balanced by the absorption of carbon dioxide by plants. The system is in equilibrium – or would be, if it were not for human interference.

The main human activity responsible for overturning the balance of the carbon cycle is the burning of fossil fuels, which adds a further 5 billion tonnes of carbon to the atmosphere over and above the natural flux each year. The main use of the energy derived is in buildings. In addition, when forests are converted to crop land, the carbon in the vegetation is oxidized through burning and decomposition. Soil cultivation and erosion add further carbon dioxide to the atmosphere.

If fossil fuels are burnt and vegetation continues to be destroyed at the present rate, the proportion of carbon dioxide in the atmosphere will treble by 2100. Even if there is decisive action on the global scale to reduce carbon emissions, atmospheric concentrations will still double by this date.

With the present fuel mix, every kilowatt hour of electricity used in the UK releases 1 kilogram of carbon dioxide. The burning of 1 hectare of forest gives off between 300 and 700 tonnes of carbon dioxide. (The carbon content of a given mass of carbon dioxide can be calculated by multiplying the mass of CO_2 by the ratio of their molecular masses – 12/44, or 0.273.)

These are some of the factors which account for the serious imbalance within the carbon cycle which is forcing the pace of the greenhouse effect which, in turn, is pushing up global temperatures.

The greenhouse effect

A variety of gases combine to form a kind of canopy over the earth which causes some solar radiation to be reflected back from the atmosphere, thus warming the earth's surface, hence the greenhouse analogy. The greenhouse effect is caused by solar heat which has been absorbed by the earth being re-emitted as long-wave radiation. This radiation is then reflected back by trace gases in the cooler upper atmosphere, thus causing additional warming of the earth's surface (see Figure 1.1, page 000). The main greenhouse gases are: water vapour, carbon dioxide, methane, nitrous oxide and tropospheric ozone (the troposphere is the lowest 10–15 kilometres of the atmosphere).

The sun provides the energy which drives weather and climate. Of the solar radiation which reaches the earth, one third is reflected back into space and the remainder is absorbed by the land, biota, oceans, ice caps and the atmosphere. Under natural conditions, the solar energy absorbed by these features is balanced by radiation from the earth and atmosphere. This terrestrial radiation, in the form of long-wave infra-red energy, is determined by the temperature of the earth/atmosphere system. The balance between radiation and absorption can change due to natural causes, such as the eleven-year solar cycle. Without the greenhouse shield, the earth would be 33°C cooler, with obvious consequences for life on the planet.

Since the Industrial Revolution, the combustion of fossil fuels and deforestation has resulted in an increase of 26 per cent in carbon dioxide concentrations in the atmosphere. In addition, the increasing population in the less developed countries has led to a doubling of methane emissions from rice fields, cattle and the burning of biomass. Methane is a much more powerful greenhouse gas than carbon dioxide. Nitrous oxide emissions have increased by 8 per cent since pre-industrial times (IPCC, 1992). All these factors contribute to the overall effect.

The state of the evidence for global warming and climate change

A statement by the Scientific Committee of the IPCC will set the stage for this discussion:

> We are certain of the following: emissions [of carbon dioxide] resulting from human activities will result, on average, in a warming of the Earth's surface. (IPCC, 1990, p.xi)

Over the last 15 years, there has been a strong world-wide warming trend. Even so, the average global temperature has risen less than models predicted. Much of the reason for this can be attributed to the Mount Pinatubo eruption, following which, the earth cooled by 0.3–0.4°C due to the shielding effect of the atmospheric debris – this just about balanced the rate of global warming. Already, the effects of the Mount Pinatubo eruption are fading, and over the next few years the world's temperatures should rapidly make up lost ground. According to James Hansen, the US climatologist: 'We predict a rapid recovery for Pinatubo cooling and new record temperatures over the 1990s' (*New Scientist*, 19 June 1993, p.7).

However, it is becoming clear that global average temperatures only tell part of the story. Climate change will not be distributed evenly across the planet. Some areas may experience much more extreme effects than others. For example, ocean currents may change direction, with profound regional consequences for climate. If the Gulf Stream was eroded or diverted, the maritime areas of northern Europe would experience significant cooling. Such a possibility has recently been endorsed by the Institute of Maritime Sciences at the University of Kiel.

The Gulf Stream is the product of a deep ocean current driven by the formation of ice in the North Atlantic, generally known as the 'deep ocean conveyor belt', which leaves behind dense, saline water which falls to the ocean bed. This deep water travels around the world before resurfacing as the Gulf Stream. This 'conveyor' is said to be unstable, and any disturbance could have a catastrophic effect, perhaps causing a 5°C cooling of average temperatures – the level attained during the depths of the last ice age. The fear is that if surface waters become warmer through global warming, less ice will be formed and the water will become fresher and less dense, thus upsetting the mechanism of the conveyor.

The historical relationship between carbon dioxide concentrations and temperature has been clearly revealed by air samples trapped in ice cores, which can be used to map the climate over the last 160,000 years. The graph showing this close correlation (see Figure 1.2, page 9) is based on data published in *Nature* (Lorius et al., 1990, p.139); it has now become part of the iconography of climatologists.

In addition, changes might occur gradually or catastrophically. This view has been reinforced by the latest batch of ice core samples. They suggest that during the interglacial period beginning 160,000 years ago, temperatures were significantly higher than at present. These higher temperatures produced severe climate perturbations, often over very short periods. This is consistent with the scientific predictions regarding the consequences of global warming.

The historical records show that when the average global temperature rose between 1–3°C, there were significant changes in the earth's climate, sea level and forest cover. There is growing acceptance of the probability that when changes due to global warming do occur, they will often be of a catastrophic nature. Such sudden changes could cause a major breakdown of ecosystems owing to imbalances between interdependent flora and fauna which have different adaptation or die-off rates. This would have chaotic consequences for sustainability, not least for humans.

Two hundred years ago, before humans began to interfere with the composition of the atmosphere in any serious way, there were about 590 billion tonnes of carbon in the atmosphere in the form of carbon dioxide. Now there are 760 billion tonnes, with most of that increase occurring over the last 50 years. This change has resulted in an increase of 1 per cent in the total solar energy absorbed by the earth.

The 1992 IPCC Report (Houghton et al., 1992) concludes that the climate consequences of a doubling of atmospheric carbon dioxide concentrations would lead to a possible rise in global average temperatures of as much as 4.5°C. The earth seems well on course to achieve this. Under its 'business as usual' scenario, the IPCC estimates that atmospheric concentrations of carbon dioxide will increase from the present 350 parts per million by volume (ppmv) to over 800 ppmv by 2100. This translates to 1.5 trillion tonnes of carbon in the atmosphere. If all known sources of fossil fuel are burnt, this figure will eventually rise to 4 trillion tonnes. These figures assume that the oceans and forests will continue to absorb about 40 per cent of all carbon dioxide emissions.

A report by the International Project for Sustainable Energy Paths (IPSEP, 1989) makes the point that climate changes would accelerate once the average global temperature had risen above 2°C, due to the action of mutually-reinforcing positive feedback systems. The earth would enter the phase of system-maximization, whereby the positive feedback loops would have increasing impact until a new level of equilibrium was reached with environmental conditions which might be inimical to human existence. The levels of carbon dioxide now in the atmosphere commit the planet to a temperature increase of 1–1.5°C. That is unavoidable. What is not certain is the rate at which this warming will occur, but the evidence suggests that it will be sooner rather than later.

One of the most worrying aspects of global warming due to human activity is that it could trigger a runaway release into the atmosphere of natural stores of carbon. Vast amounts of carbon dioxide are trapped in tundra ice and are being released as melting occurs. It is known that permafrost also traps large amounts of methane and other organic gases. Until recently, it was thought that these reservoirs of gases occurred at a minimum depth of 200 m. A recent core sample taken from the Mackenzie River in Canada revealed that large stores of methane exist at 50–70 m below ground. This means that these stores of methane are much more vulnerable to thawing than was previously thought. As methane is a potent greenhouse gas, this threatens to increase the pace of global warming.

Another factor to consider is that as oceans warm, they release carbon dioxide. At the same time, the carbon-fixing capacity of the planet is being reduced by uncontrolled deforestation and the death of trees because of pollution and excessive UVB radiation due to ozone depletion.

Some of the most convincing evidence of biological change due to global warming has come from the British Antarctic Survey. Retreating ice sheets have led to increases in plant growth in certain species up to 25-fold. This is attributed to the fact that average temperatures have risen 1°C over the last 25 years ('Antarctic warning', *New Scientist*, 25 June 1994, p.11), and, even more importantly, Antarctic summers have lengthened by 50 per cent since the 1970s, giving plants a much better chance to reproduce. Also, new species are appearing as glaciers retreat. Spores of plants blown on the winds which have lain frozen in the ice perhaps for centuries are now coming into growth. Perhaps this could be the 'hard' evidence the sceptics keep demanding.

Why this focus on carbon dioxide? Emissions of this gas from the burning of fossil fuels contribute more than half of the total potential warming from all greenhouse gases, therefore it is 'the single most important driving force behind the threat of greenhouse warming' (IPSEP, 1993). Furthermore, carbon dioxide and several other greenhouse gases have a long lifetime within the atmosphere, so once the momentum of global warming has been established, it will continue for a considerable time. Even if there was an immediate cessation of all greenhouse gas emissions, this would not stop the process. The planet is committed to increasingly detrimental climate change, regardless of future policies. The area of choice is confined to the amount of damage nations and their people are prepared to tolerate.

The IPCC states that the current target adopted by the industrialized countries of returning carbon emissions to 1990 levels by 2000 will still result in rising concentrations for at least a century. Emissions of carbon dioxide would have to fall steeply and quickly for the atmosphere to stabilize at present-day levels. However, emissions would have to fall well below 1990 levels in the next century for the atmosphere to stabilize even at twice the pre-industrial level. This would mean that average global temperatures would still rise substantially. There is also a basic flaw in the UN Framework agreement, in that it sets in stone the enormous differences in carbon emissions between countries.

To put the UN Framework Agreement into perspective, the IPCC Scientific Committee recommends that carbon emissions world-wide should be cut by 60 per cent compared to 1990 levels if there is to be a reasonable chance of arresting global warming. Another recent conclusion is that methane emissions may have twice the warming capacity previously thought, due to the fact that methane depletes the atmospheric chemicals that break it down.

One of the greatest uncertainties in attempting climate modelling is the effect of feedback loops. For example, methane emissions from natural wetlands and rice paddy fields are significantly larger at higher temperatures. Methane is one of the most powerful greenhouse gases; consequently, the more methane, the greater the warming, leading to even greater emissions from wetlands and paddy fields. According to the IPCC, the net effect of feedback systems will be to intensify greenhouse warming. It states: 'For this reason, climate change is likely to be greater than the estimate we have given' (IPCC, 1990, p.xxvii).

The UN IPCC world wide network of 2,500 scientists, meeting at Maastricht in September 1994, confirmed that the conclusions it reached in 1990 and 1992 'do not substantially change'. This is despite the impact of the Mount Pinatubo eruption and the increased knowledge about the impact of sulphate aerosols – one of the main by-products of the combustion of fossil fuels – on the atmosphere. The latter may be reducing global warming by as much as a half, but this will not prevent climate change, since aerosols are unevenly mixed across the globe. They may even add to the problems associated with climate change, since uneven degrees of warming will probably alter the movement of air masses.

The 1995 IPCC update report (Metorological Office, 1995) asserts that there is virtually no room for doubt that human activity is causing the planet to warm. It bases this belief, amongst other things, on the fact that its earlier predictions about the rise in mean global temperature and the rate of natural disasters have all been realized. The report reiterates that the effects of past emissions of the main greenhouse gases will persist for centuries, even if drastic curbs are introduced immediately. Scientists are now more concerned than ever that the failure of politicians to achieve radical carbon abatement policies could result in runaway global warming.

The General Circulation Model

The predictions concerning global temperature changes due to the accumulation of greenhouse gases in the upper atmosphere have been produced by general circulation models (GCMs) run by powerful supercomputers. They have achieved complex representations of the earth's atmosphere and, by simulating changes in the chemical composition of the atmosphere to reflect human activity, they have predicted the probable changes in the earth's climate.

One of the main arguments employed by the sceptics is that there is no way of verifying the validity of such models. This would be true if there were no GCM which could accurately reproduce *past* fluctuations in the earth's temperature on the basis of previous data. However, now there is such a model, and it has successfully predicted the temperature events which occurred since records began in 1860. This means its future predictions can be treated with much greater confidence than was the case with previous GCMs. This is a

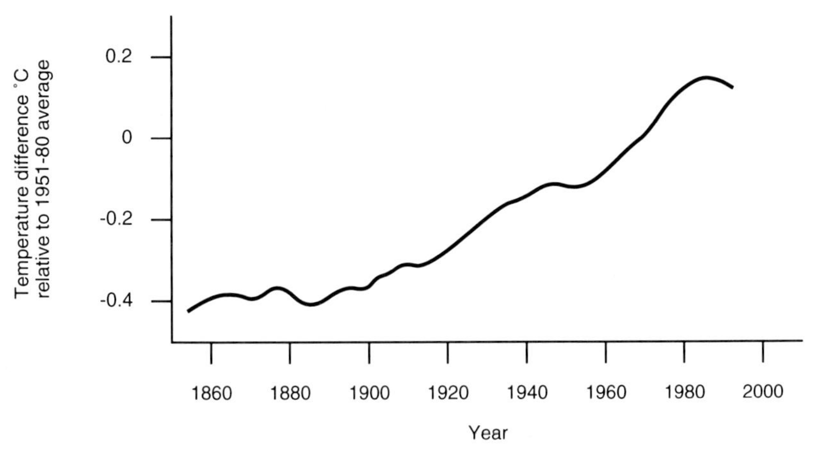

Figure A.1 Change in global temperature since 1880

considerable achievement on the part of the Hadley Centre for Climate Predictions and Research at the UK Meteorological Office.

Much of this accuracy is due to the fact that the Hadley model takes account of the pollution caused by sulphate particles emitted during the burning of fossil fuels. As stated earlier, it has been known for some time that these particles are having a dampening effect on global warming by reflecting solar radiation back into space. The Hadley team estimates that this effect is reducing greenhouse warming by about 30 per cent (DoE, 1994b; David Carson, Director, Hadley Centre for Climate Prediction, *Royal Society of Arts Journal*, June 1996).

Despite the impact of sulphate particles, the net effect, according to Hadley, is that global warming is occurring 'twice as fast as the level at which we can expect to see significant impacts on the biosphere' (David Bennetts, Hadley Research Co-ordinator, *New Scientist*, 26 November 1994, p.6). In other words, we cannot rely on pollution to prevent climate damage from global warming. The correlation between human activity in the form of industrial performance, and global warming is illustrated by the delayed but steep emissions of carbon dioxide and temperature rises which occurred following the two world wars. Figure A.1 illustrates this relationship.

The Hadley team endorses the IPCC scientists by concluding that the world is warming at an unprecedented rate, and that this is almost wholly due to human activity. Supporting evidence has come from the US government's Climate Prediction Centre, which has revealed that average global temperatures between March and October 1994 were 0.4°C above normal. This is a considerable rise in climatic terms, and makes it the hottest period since records began. Similar results have been obtained by the University of East Anglia Climate Research Unit, which confirms the prediction that temperatures would rise once the debris from the Mount Pinatubo eruption had settled to earth. Temperature records for 1994 show that it was the third hottest year on record – it would have been the hottest but for exceptionally low temperatures in January and February. Following this, 1995 has been confirmed as *the* hottest year on record.

Global warming: A litmus test

Whilst governments may prevaricate over the seriousness of the threats arising from global warming, this is not the case with the international insurance market – how does it react to the phenomenon of global warming?

Catastrophe insurance operates by spreading the risk across a network of insurers and reinsurers. Increasingly, individual members of the network are unable to bear their share of the burden because of the mounting incidence of claims arising from storm damage.

Following the 1990 storms, the Swiss insurance company Swiss Re stated:

> There is a significant body of scientific evidence indicating that last year's record insured losses from natural catastrophes was ... the result of climatic changes that will enormously expand the liability of the property-casualty industry. In the light of the magnitude of these losses, it would be prudent to act as if [global warming] is correct. Failure to act would leave the industry and its policyholders vulnerable to truly disastrous consequences. ('Nature takes its revenge', *Independent on Sunday*, 1 May 1994)

The reaction to the 1990 storms by another major company, Munich Re, was to state:

> For the first time in the history of our planet, mankind is about to change the climate significantly and possibly irreversibly, without having any idea of the consequences that will have. ('Storms get worse as world warms up', *Guardian*, 13 September 1994)

As if to confirm this statement, the worst floods for 150 years occurred in 1993 along the Mississippi and Missouri rivers, costing £8 billion; in December 1993, the worst floods for over 60 years hit northern Europe, and in November 1994, floods throughout the River Po valley had a devastating impact on life and property. In northern Italy, the frequency of major floods has increased by 50 per cent over the last two decades. The conclusion of Munich Re: 'It is noticeable and worrying that for the last six years nearly every year has brought a new record storm loss.'

What is concentrating the minds of insurers is the probability of increasing damage to property in highly-insured, densely-populated areas. The largest insurance companies are now employing their own climatologists. One of these, Gerhard Berz, head of the Technical Research Division of Munich Re, has said: 'The increased intensity of all convective processes in the atmosphere will force up the frequency and severity of tropical cyclones, tornadoes, hailstorms, floods and storm surges in many parts of the world, with serious consequences for all types of property insurance.'

The danger zone includes the coasts of western Europe. The ultimate nightmare is the 'supercyclone', with sustained wind speeds in excess of 240 kilometres per hour. If Manhattan experienced such an event, it would kill the insurance industry. This is not improbable, because as seas warm, so the critical 26°C above which storms occur will be experienced across ever wider latitudes.

After Hurricanes Andrew and Iniki, nine American insurance companies collapsed. Yet things could have been worse: Hurricane Andrew caused $15 billion of damage; but if it had veered 20 miles north to hit downtown Miami, the insurance bill would have reached $75 billion. Another slight deviation could have taken it to New Orleans, with an estimated overall cost of $100 billion. In October 1995, Hurricane Opal hit Florida, with gusts reaching 180 m.p.h. In August that year, Dr William Gray, leader of a group of atmosphere scientists at Colorado State University, used a sophisticated model to predict that there would be an increase in major storms hitting the USA. As Gray stated: 'We are going to see hurricane damage like we've never seen it before' ('Hard wind a'blowing', *Guardian*, 13 September 1995), which is why insurance companies are now speculating on the possibility that global warming-related claims will exceed the total resources of the industry, currently standing at $1.3 trillion.

Insurers may well be interpreting the shape of events to come by considering the massive floods in China which occurred in July 1995. The river system of the Hubei and Hunan provinces dominated by the Yangzi recorded the highest levels since records began 125 years ago. This is said to be due to a combination of deforestation and an exceptional rate of melting of snows on the high Tibetan plateau, which is attributed to global warming. Ironically, large areas of northern China were simultaneously affected by severe drought, reinforcing the prediction by IPCC scientists that global warming would produce greater extremes of climate.

The final twist to the argument is provided by evidence emerging from a recent core drilled by the Greenland Ice Core Project (GRIP). Previously, climatologists had thought that the last warm interglacial period, which ended 114,000 years ago, was relatively stable. This latest evidence shows that a warming of the planet produced periods of considerable instability, with sharp changes of temperature occurring over very short timescales. This was mainly evident in the northern hemisphere, probably due to the North Atlantic current 'switching on and off' and the movement of icebergs. This interglacial period ended abruptly, with temperatures in Greenland dropping 14°C in ten years.

It seems that the build-up of greenhouse gases is directly linked to the steep fluctuations in climate that are a characteristic of this period in the northern hemisphere. The effects of the pollution of the atmosphere could well be much more dangerous than previously thought, causing dramatic changes that are much too abrupt to accommodate natural rates of adaptation.

Predicted changes in climate arising from global warming

Earlier sections in this Annex examined the evidence for global climate change. Now we turn our attention to the likely changes to the general climate, and the effects on the environment and buildings.

If the nations of the world, especially the industrialized countries, maintain 'business as usual' until the end of the next century, the IPCC Scientific Committee predicts that the average global temperature will have risen by over 4°C, and possibly by as much as 6°C or more. There are numerous possible consequences of such an increase. Higher temperatures and more extreme weather conditions will have significant effects on buildings, and thus on the way in which they will need to be designed and serviced. In this Annex, it will only be possible to summarize some of the probable outcomes of the changing climate. The discussion begins with one of the most dramatic – a significant rise in sea level.

Sea level

Two factors are causing the level of the seas to rise: the melting of glaciers and thermal expansion. The Antarctic and Greenland ice sheets are also making a contribution, but their effects are unpredictable, which inserts a major element of uncertainty into the reckoning.

Under 'business as usual' conditions, the IPCC best-case estimate is that by 2100, sea levels will have risen by almost 70 cm, with the worst-case scenario reaching 110 cm. By 2030, the level will have risen by almost 20 cm. Thereafter, even if carbon dioxide emissions were stabilized at the 2030 level, the momentum in the system would cause sea levels to rise for decades, perhaps even centuries. By 2100, under such conditions, the level would have risen to over 40 cm, and would continue rising. Greatest concern is centred on the West Antarctic ice sheet, much of which is grounded far below sea level. Its volume is equivalent to 5 m of global sea level, and if there were to be a sudden outflow of ice due to higher sea temperatures, this would have catastrophic results.

In summary, rising sea levels would have a number of consequences:

- the erosion of coasts;
- flooding of unprotected low-lying areas;
- a shift in wetlands;
- increased risk of storm surges;
- extended salination of estuaries;
- threats to drinking water supplies near coasts;
- altered tidal rise and fall;
- changes in the rate of transport of sediment
- serious consequences for buildings already erected at or below sea-level;
- the need to redesign buildings for low-lying regions.

The rise in sea level would be unevenly distributed for two main reasons.

Thermal expansion: Changes in ocean currents and surface air pressure will vary from region to region. Assuming a given temperature rise, varying ocean depth and therefore differing rates of mixing and consequent expansion will lead to considerable differences in sea level rise. Initial modelling suggests that north-west Europe would experience rises considerably above average, whilst levels in Antarctica would actually fall.

Vertical land movement: This can produce changes in the relative levels of land and sea equally as severe as the warming effect. For example, the south-east of England is tilting downwards, which will add considerably to its problems. Louisiana is losing 250 km^2 a year through tectonic subsidence.

Whole countries and several major cities are directly threatened by rising sea levels. In India, 5,700 km^2 of coastal area is at risk of inundation. Over 7 million people could be displaced, as is also the case in Vietnam. Bangladesh is already frequently subject to catastrophic floods. Islands world-wide will be at risk, as will major cities like Singapore. Even London will have its problems, since the Thames Barrage is already acknowledged to offer inadequate protection within this scenario. Many of the intensively-farmed areas of the industrialized countries are at risk, including the coastal regions of the North Sea, the estuary of the River Po, Louisiana and the Gulf of Finland. Altogether, it is estimated that 50 per cent of the world's population lives on or near coasts.

The 1 m rise in sea level which seems almost inevitable will result in massive migrations of population. UNEP estimates that 50 million people would be displaced by a sea level rise of 1.5 m. As far as insurance companies are concerned, coastal land below the 5 m contour line is now regarded as posing a special risk, with the result that buildings in such areas already bear higher premiums. In eastern England, large areas lie below this level. This, coupled with the fact that drainage has led to land sinking by as much as 3 m in the Fens near Peterborough, places this area at particular risk. Figure A.2 shows height contours for England and Wales, and demonstrates the size of the areas of concern. Similar maps for a number of other countries could be produced to illustrate similar or worse positions.

One of the major areas of uncertainty concerns the land-based Antarctic ice sheet, which covers an area greater than Europe and contains 90 per cent of all fresh water on the planet. If it were to melt completely, the sea level world-wide would rise by over 60 m. In September 1993, a conference of glaciologists from 22 countries held in Cambridge, UK, concluded that this outcome is extremely unlikely, but that smaller rates of break-up are possible, and would have a catastrophic impact on maritime countries. Evidence from the Norwegian Polar Research Institute suggests that the ice shelves may be breaking up far faster than expected. Consequently, a sea level rise of up to 5 m is possible, and it would occur over a relatively short period.

In Autumn 1995, the Nansen Environmental and Remote Sensing Centre obtained evidence that sea ice around Antarctica was melting. The justification for this assertion comes from collating the data from two satellites: Nimbus 7, operating between 1978 and 1987, and the US Defense Meteorological Satellite Programme from 1987. This reveals a statistically significant decrease in sea ice of 1.4 per cent per decade, which accords with data from the Australian Antarctic Division in Hobart. This, coupled with the break-up of two massive ice shelves, is directly in line with the predicted effects of global warming. In fact, the average temperature in Antarctica has risen 2.5°C over the past 50 years, which is a considerable increase.

There is one final speculation connected with rising sea levels. There is evidence that there is increased volcanic activity when seas are either falling or rising. More volcanoes seem to erupt explosively with rising sea levels. This could be due, on the one hand, to the erosion of coastal volcanoes and volcanic islands. This was said to be the cause of the Mount St Helens eruption. On

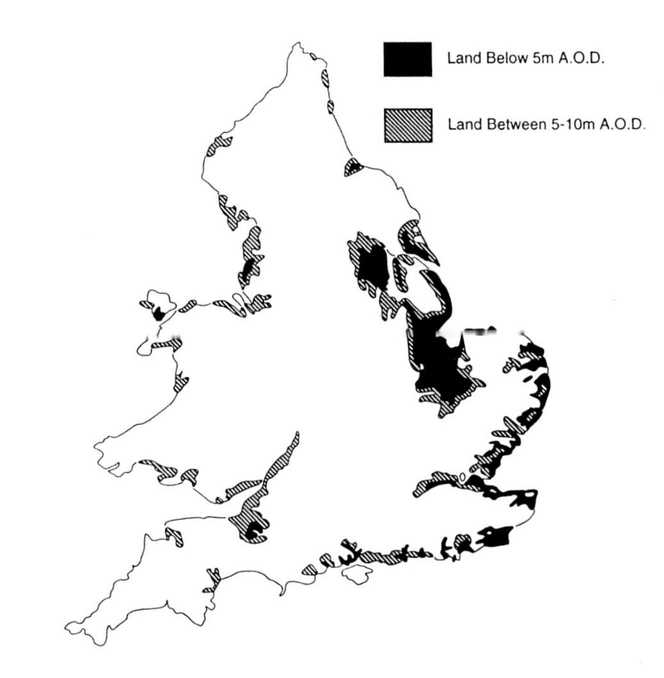

Figure A.2 Land below the 5 and 10 m contours in England and Wales (AOD = Above Ordnance Datum)

the other hand, there is the chance that the increased weight of water bearing down on volcanoes may cause faults to open up, allowing fresh magma to rise to the surface. Some of the most violent past eruptions, like Krakatoa in Java, resulted from magma coming in contact with water.

To add to these problems, the effects of these predicted rises in sea level may be greatly enhanced by storm surges. Scientists in the University of East Anglia School of Environmental Sciences published a study in 1992 which predicted the effect of rising sea level and sinking land on coastal towns in the UK (Barkham et al., 1992). Their yardstick was the National Rivers Authority standard of the 1 in 100-year storm. They compiled a table ot predictions for the 1 in 100-year storm for 2030 and 2100. By 2030, Milford Haven, Cardiff, Portland, Newhaven and Colchester would all experience such storms every five years or less; for Aberdeen, the figure was every six years, and for Southampton, every eight years. By 2100, most storm return rates were measured in months.

The reason for this alarming prognosis is that global warming not only causes sea levels to rise, it also increases the ferocity of storms. The intensity of low-pressure systems is increased, which not only leads to higher winds, but also causes the sea level to rise locally due to the lowering of surface pressure. The rise in level can be as much as 60 cm.

A statistic which reinforces these predictions has come from British oceanographers, who have discovered that, over the past 30 years, average wave height in the North Atlantic has risen by up to 50 per cent ('Above us the waves', *Guardian*, 4 May 1995). In the USA, Oceanweather Inc. of Connecticut has found that there has been a dramatic increase in the size and incidence of 'monster' waves – waves almost twice as high as those normally seen in hurricanes and 50 per cent larger than the so-called 1 in 100-year waves ('Above us the waves', *Guardian*, 4 May 1995). They have occurred twice off the coast of Canada in two years. Satellite evidence reveals that sea levels are rising at twice the predicted rate. Half a million readings per day by the Topex/Poseidon satellite indicate that seas have risen 3 mm per year in two years (DoE, 1994a), which is double the best-case estimate by the IPCC scientists.

There is one final contributory cause to rising sea levels which has nothing to do with global warming but *is* a consequence of human activity. The draining of wetlands, underground aquifers and inland seas has added significantly to the level of the oceans. Prior to the Topex/Poseidon evidence, climatologists estimated that the seas would be rising at a rate of 1.5–2 mm per year. Of this, 0.5 mm could be directly due to human activity, including deforestation ('How disappearing lakes are swelling the oceans', *New Scientist*, 22 January 1994, p.17). As a classic example, the Aral Sea has lost more than 60 per cent of its volume since 1960 due to the diversion of its feeder rivers for irrigation. In 20 years, it will have disappeared, and by then it will have added 3 mm to global sea levels.

The current rate of loss of rainforests of 150,000 km^2 a year and the consequent erosion of their soils will release about 50 km^3 of water per year, adding a further 0.14 mm to the level of the oceans ((Lorius et al., 1990)

Rainfall, vegetation and desertification

As temperatures rise, the level of water vapour in the atmosphere will increase considerably. Also, with each degree of temperature rise, global mean precipitation will increase by 2–3 per cent per year. Where and by how much the changes in rainfall will occur are still uncertain, given the limitations of current climate models. On the basis of available research, the German Enquete Commission concludes: 'Both extreme values of the water cycle will increase, which means particularly heavy precipitations ...' (Enquete Commission, 1991).

Heavier rainfall will, in many cases, be coupled with deeper low-pressure systems due to more vigorous convective dynamics. Climate change models predict a decline in soil humidity in continental inland areas, placing additional stress on already overstretched aquifers. Millions of hectares of agricultural land in the USA have been abandoned because ground water reserves are exhausted. The cereal-growing belt of the Great Plains is at risk as the costs of irrigation mount. The spectre of the 'dust bowl' phenomenon of the 1930s is haunting the whole region, since, in Arkansas for example, the rains have failed for three successive years (1994–6). The demand for ground water for irrigation has risen dramatically over the century. Since 1900, the amount of land irrigated for agriculture has risen from 40 million to 257 million hectares.

In places where food production is already marginal, very small changes in temperature and rainfall can have catastrophic effects, as has already been demonstrated in the Horn of Africa. Such places include Maghreb, western and southern Africa,western Arabia, south-east Asia, Mexico, central America and eastern Brazil. On the other hand, temperature rises in the high northern latitudes could mean that the tundra is pushed north by hundreds of kilometres.

Until recently, it was thought that the tropics would be the region least affected by global warming. This view has had to be revised, with some arguing that a rise in temperature in the tropics could be devastating. The main reason for this opinion is that, as land and air temperatures increase, so the atmosphere can hold more moisture. In the tropics, an increase of 4°C in the air temperature would mean that about 30 per cent more water evaporating from the ground could be held in the atmosphere. The net effect of this would be to cause land to dry out, especially at low and subtropical latitudes. Arid conditions would extend about 35° north and south of the equator. By 2060, it is expected that up to 350 million people will be affected by famine due to climate change. This is on top of the baseline population perennially at risk from hunger which will have risen to 640 million due to population growth. According to the UK Meteorological Office model, if the temperature rose by 5°C, the number at risk due to famine would be in the order of 750 million.

A recent study has examined the consequences for world grain yields by 2060, assuming a mean global temperature rise of 4°C. At this temperature, yield changes ranged from +30 per cent to -30 per cent. Where crop yields are improved, this is due to the increased level of atmospheric carbon dioxide. However, increases are dependent on the maintenance of adequate water supplies – a situation far from certain given the predicted rise in water stress in the most vulnerable parts of the world. A UK Meteorological Office study concludes that if farming on a global level continues with business as usual, ignoring climate change, the affect on agriculture, and consequently food supplies, could be drastic. Overall, a temperature increase of the order of 4°C would cause a 10 per cent decline in the production of wheat, maize, soya beans and rice in developing countries (Parry and Rosenzsweig, 1994). At the same time, with no serious attempts to adapt to global warming, the UK model predicts an increase in the cereal price index of almost 150 per cent.

Pests and pathogens

Global warming will bring about changes in the distribution of pests and diseases. According to the US Environmental Protection Agency, diseases currently confined to the tropics, like swine fever, will reach the USA, causing severe economic damage. There has already been an example of pest migration, with the locust invasion of southern Europe in 1986–8 reaching a new northern limit. In areas where pests and pathogens already exist, their effect will become more severe as temperatures rise. The malarial mosquito is another insect which it is predicted will migrate north. The latest estimate, based on five models of the changes in temperature, rainfall and humidity resulting from a doubling of atmospheric carbon dioxide, is that seasonal malaria will spread to large parts of the UK. The IPCC estimate is that this concentration of carbon dioxide will be reached in the second half of the 21st century.

In September 1995, a medical conference in Washington, DC, considered the effects of global changes on health (reported in 'Global changes in climate will affect health', *British Medical Journal*, Vol. 311, 23 September 1995). Delegates were warned that deaths due to heat alone would increase significantly, and that the 1995 toll of 500 in Chicago was a sign of things to come. More alarming was the probable change in the distribution of diseases such as dengue fever, malaria and Lyme disease. Alterations in the operation of the El Niño Pacific current due to global warming could have an adverse effect on food supply, pollution and the availability of water.

In July 1995, scientists at the UK government's Rothamsted Agricultural Station warned farmers that there was about to be 'a plague of Biblical proportions' of aphids, both greenfly and blackfly ('Crops threatened by aphid plague of Biblical proportions', *Independent on Sunday*, 23 July 1995). A mild winter and exceptionally hot summer had led to populations 'tens or even hundreds of times the usual quantity' – by far the greatest number since records began in the 1960s. The expectation is that crops such as sugar beet and potatoes will be reduced by as much as 10 per cent due to aphid damage. At about the same time, the Medical Entomology Centre at Cambridge University reported an eruption of small black 'thunderflies' in hundreds of sites in the UK ('Crops threatened by aphid plague of Biblical proportions', *Independent on Sunday*, 23 July 1995). They flourish in hot, humid conditions.

Climate change and ecosystems

Existing models can only offer very generalized conclusions regarding the effect of climate change on ecosystems. The ecosystem processes of photosynthesis and respiration are dependent on climate factors and carbon dioxide concentrations in the short term. In the longer term, climate and carbon dioxide levels control the actual composition of ecosystems, either directly by killing off poorly-adapted species, or indirectly by influencing competition between species. The more complex an ecosystem, the more it is likely to be destabilized by a rapid change in temperature and/or a change in rainfall pattern.

Terrestrial ecosystems are subject to constant change, but at a slow rate. In addition, they have a buffer capacity which enables them to survive episodes of extreme weather. Breakdown occurs if extremes of weather last for long periods, or if they occur frequently and in rapid succession. Both these scenarios are predicted by climate change models.

As an example, recent models predict water shortages in the southern temperate forests of the northern hemisphere due to climate change. This will be a primary cause of forest death. A feedback loop will come into play, as forests are replaced by grassland, with its more limited capacity to fix carbon. As a result, there will be a net transfer of carbon to the atmosphere, which in turn will lead to greater global warming. This may lead to even more severe water shortages and even greater forest death, and so on. Forests are an example of very large-scale ecosystems which may not be able to migrate fast enough to keep pace with climate change.

An ecosystem is a highly complex and often fragile network of interactions between organisms, both plant and animal.. The loss of any one component could lead to the break-up of the whole system. This is one of the feared consequences of rapid climate change. Organisms within a stable ecosystem will probably have differing capacities to adapt to relatively abrupt changes in temperature and rainfall patterns. Some ecosystems may collapse quite early in the timescale of changing climate.

Summary of impacts

The 1990, 1992 and 1995 IPCC scientific assessments lead to the following conclusions regarding climate changes resulting from global warming:

- Continents will warm more than oceans.
- Temperature increases in southern Europe and North America will be greater than the global mean.
- In Europe and North America, there will be reduced rainfall and soil moisture in summer.
- The Asian summer monsoon will intensify.
- Episodes of high temperature will increase.
- There will be a general increase in convective precipitation.
- The effect of deforestation on climate may be significant due to the contraction of the carbon dioxide sink. Broad-leafed trees are especially effective at absorbing carbon dioxide from the atmosphere to facilitate growth. Tropical deforestation could lead to substantial local effects, including a reduction of precipitation by about 20 per cent (which is one reason why the growing practice of counteracting the carbon emitted by a building in use by funding reforestation in a distant country is to be commended).

To these predictions must be added the most serious of all likely consequences: the rise in sea level, with its implications for the loss of food-producing land and the displacement of whole populations. Whilst most adverse consequences will mainly affect the less developed countries, rising sea level will be totally equitable, and as devastating for the economies of the industrialized as the developing countries.

In Article 2 of the Climate Change Convention, the Earth Summit at Rio de Janeiro laid down the objective 'to achieve the stabilization of greenhouse gases in the atmosphere at a level that would prevent dangerous anthropogenic interference with the climate system'. Such stabilization must take place 'within a time frame to allow ecosystems to adapt naturally to climate change, to ensure that food production is not threatened and to enable economic development to proceed in a sustainable manner'. Such an objective will certainly not be achieved while the industrialized countries continue to cling to 'business as usual', particularly in their use of fossil fuels.

To repeat the World Energy Council's estimate, in 25 years, energy use will have increased by 88 per cent over the 1990 level, representing a growth rate of 3 per cent per year (WEC, 1993). Most of this energy will come from fossil fuels. More recent predictions regarding the growth rate of cities along the Pacific rim make even this look conservative. For example, a linear city is developing along 70 miles of the Pearl River in China. These hypercities are being constructed without any consideration of energy conservation.

With an estimated world population of 8 billion by 2020, the World Energy Council has predicted a 44 per cent increase in greenhouse gases by this date, driven largely by the increased use of fossil fuels by developing countries. Most recent trends show that carbon dioxide emissions from the developing countries are rising rapidly. They will soon account for one third of global emissions, and are set to double every 14 years (Brown et al., 1995, p.66).

In the face of this evidence, it might seem that the efforts of those engaged in producing buildings and creating cities can make little impression on the problem. On the contrary: the way the design or retrofitting of buildings world-wide progressively eliminates the use of fossil-based energy could make an impact on global warming exceeding anything that might be agreed by politicians.

Bibliography

Anderson, V (1993) *Energy Efficiency Policies*, Routledge, London

Anink, D, Boostra, C and Mak, J (1996) *Handbook of Sustainable Building*, James and James, London.

Baker, N V and Steemers, K. (1994) *The LT Method 2.0*, Cambridge Architectural Research, Cambridge, UK.

Barker, T (1993) 'The Economic Feasibility of Achieving a 60% Reduction in UK CO_2 Emissions by 2040', paper presented at the Symposium on the Environment and British Energy Policy, Green College, Oxford, UK, 27 April.

Barker, T, Bayliss, S and Bryden, C (1994) *Achieving the Rio Target: CO_2 Abatement Through Fiscal Policy in the UK*, Energy-Environment-Economy Modelling Paper No. 9, June, Department of Applied Economics, University of Cambridge, UK.

Barkham, J P, Macguire, F A S and Jones, S J (1992) *Sea Level Rise and the UK*, Friends of the Earth, London.

Bell, J and Burt, W (1995) *Designing Buildings for Daylight*, Building Research Establishment Report BR 288, Construction Research Communications Ltd, Watford, UK.

Bordass, W, et al. (1991) 'Daylight in Open-plan Offices: The Opportunities and the Fantasies', paper presented to CIBSE National Lighting Conference, UK.

BRE (1994) *Thermal Insulation: Avoiding Risks* (2nd edn), Building Research Establishment and HMSO, London.

BRECSU (1991) *Energy Efficiency in Offices*, Energy Consumption Guide 19, Building Research Energy Conservation Support Unit, Energy Efficiency Office, Department of the Environment, HMSO, London.

BRECSU (1993) *Energy Efficiency in New Housing*, Good Practice Guide 79, Building Research Energy Conservation Support Unit, Energy Efficiency Office, Department of the Environment, HMSO, London.

BRECSU (1995) *A Performance Specification for the Energy Efficient Office of the Future*, General Information Report 30, Building Research Energy Conservation Support Unit, Energy Efficiency Office, Department of the Environment, HMSO, London.

Brown, L R, Lenssen, N and Kane, H (1995) *Vital Signs 1995–1996*, Worldwatch Institute, Earthscan Publications, London.

Button, D and Pye, B (eds) (1993) *Glass in Building*, Butterworth Architecture, London.

Cofaigh, E O, Olley, J A and Lewis, J O (1996) *The Climatic Dwelling*, James and James, London, on behalf of the European Commission Directorate-General for Energy.

DGXVII (1992) *Energy Efficient Lighting in Buildings*, Thermie Programme Action Maxibrochure, Rational Use of Energy and Renewable Energy Programmes, European Commission Directorate-General for Energy, Brussels, Belgium.

DGXVII (1994a) *Natural and Low Energy Cooling of Buildings*, Thermie Programme Action Maxibrochure, Rational Use of Energy and Renewable Energy Programmes, European Commission Directorate-General for Energy, Brussels, Belgium.

DGXVII (1994b) *Daylighting in Buildings*, Thermie Programme Action Maxibrochure, Rational Use of Energy and Renewable Energy Programmes, European Commission Directorate-General for Energy, Brussels, Belgium.

DGXVII (1995) *Environmental Assessment of Buildings*, Thermie Programme Action Maxibrochure, Rational Use of Energy and Renewable Energy Programmes, European Commission Directorate-General for Energy, Brussels, Belgium.

DoE (1994a) 'Satellite programme to speed detection of climate change', *Climate Change*, No. 2, Spring, Department of the Environment, London.

DoE (1994b) *Climate Change: Science Update*, Autumn, Department of the Environment, London.

DoE (1995) *The Building Regulations Approved Document L: Conservation of Fuel and Power*, HMSO, London.

Edwards, B (1996) *Towards Sustainable Architecture*, Butterworth Architecture, London.

Enquete Commission (1991) 'Extreme Weather Events', *Protecting the Earth: Third Report of the Enquete Commission of the 11th German Bundestag*, Bundestag, Bonn, Germany, p.200.

ETSU (1993) *House Design Studies: Overview*, ETSU S 1362, Energy Technology Support Unit, Department of Trade and Industry, London.

Goulding, J R, Lewis, J O and Steemers, T C (1992a) *Energy Conscious Design: A Primer for Architects*, B T Batsford, London, for the Commission of the European Communities.

Goulding, J R, Lewis, J O and Steemers, T C (1992b) *Energy in Architecture: The European Passive Solar Handbook*, B T Batsford, London, for the Commission of the European Communities.

Grubb, M (1990) *Energy Policy and the Greenhouse Effect*, Vol. 1, Royal Institute of International Affairs, London.

Hawkes, D (1996) *The Environmental Tradition*, E & F N Spon, London.

Houghton, J T, Jenkins, G J and Ephraums, J J (eds) (1990) *Climate Change: The IPCC Scientific Assessment*, Cambridge University Press, Cambridge, UK.

Houghton, J T, Callander, B A and Varney, S K (eds) (1992) *The Supplementary Report to the IPCC Assessment*, Cambridge University Press, Cambridge, UK.

IEA (1996) *Energy in Buildings and Industry*, International Energy Association, May.

IPSEP (1989) *Energy Policy in the Greenhouse*, Vol. 1, International Project for Sustainable Energy Paths, El Carrito, California.

IPSEP (1993) *Energy Policy in the Greenhouse*, Vol. 2, International Project for Sustainable Energy Paths, El Carrito, California.

Littlefair, P J (1989) *Innovative Daylighting Systems*, Building Research Establishment Information Paper IP 22/89, BRE, Watford, UK.

Lorius, C, Jouzel, J, Reynaud, D, Hansen, J and Le Treut, H (1990) 'Greenhouse Warming, Climate Sensitivity and Ice Core Data', *Nature*, Vol. 347, September.

Meteorological Office (1995) *Climate Change 1995: The Science of Climate Change – Summary for Policy Makers*, Meteorological Office Graphics Studio 95/869, London.

National Audit Office (1994) *Buildings and the Environment,* HMSO, London.

Nelson, G (1995) *The Architecture of Building Services*, B T Batsford, London.

Olivier, D and Willoughby, J (1996) *Review of Ultra-low-energy Homes*, General Information Report 38, Energy Efficiency Office/BRECSU, HMSO, London.

Parry, M. L. and Rosenzsweig, C. (1994) 'Potential Impact of Climate Change on World Food Supply', *Nature*, Vol. 367, pp.133–8.

Pearce, D (1993) *Blueprint 3: Measuring Sustainable Development*, Earthscan Publications, London.

Photovoltaics in Buildings Group (1995) *Photovoltaics in Buildings: Technical Information Pack*, IT Power, Hampshire.

Romm, J J and Browning, W D (1995) *Greening and the Bottom Line*, Rocky Mountain Institute, California.

Schipper, L and Meyers, S (1992) *Energy Efficiency and Human Activity: Past Trends and Future Prospects*, Cambridge University Press, Cambridge, UK.

Smith, P F (1996) *Options for a Flexible Planet*, Sustainable Building Network, University of Sheffield, UK.

Thomas, R (1996) *Environmental Design*, E & F N Spon, London.

Vale, B and Vale, R (1991) *Green Architecture*, Thames and Hudson, London.

WCED (1987) *Our Common Future* (The Brundtland Report), Oxford University Press, Oxford, UK, for the World Commission on Environment and Development.

WEC (1993) *Energy for Tomorrow's World*, World Energy Council, St Martin's Press, London.

Yannas, S (1994) *Solar Energy and Housing Design*, Vols 1 and 2, The Architectural Association, London.

Index

This index is compiled on a word by word basis so that for example *New Scientist* comes before Newcastle. Bold location references are illustrations, italicised location references are diagrams.